Cotton Fiber Chemistry and Technology

INTERNATIONAL FIBER SCIENCE
AND TECHNOLOGY SERIES

Series Editor

MENACHEM LEWIN
Hebrew University of Jerusalem
Jerusalem, Israel
Herman F. Mark Polymer Research Institute
Polytechnic University
Brooklyn, New York

The Editor and Publisher gratefully acknowledge
the past contributions of our distinguished
Editorial Advisory Board

INTERNATIONAL FIBER SCIENCE AND TECHNOLOGY SERIES

Series Editor: **MENACHEM LEWIN**

International Fiber Science and Technology

Cotton Fiber Chemistry and Technology

Phillip J. Wakelyn

Noelie R. Bertoniere

Alfred Dexter French

Devron P. Thibodeaux

Barbara A. Triplett

Marie-Alice Rousselle

Wilton R. Goynes, Jr.

J. Vincent Edwards

Lawrance Hunter

David D. McAlister

Gary R. Gamble

CRC Press

Taylor & Francis Group

Boca Raton London New York

CRC Press is an imprint of the
Taylor & Francis Group, an **informa** business

Preface

This was published originally as a chapter in the *Handbook of Fiber Chemistry, Third Edition*. Because of particular interest in the new revised and expanded Cotton Fiber chapter, it is also now being published as a separate book. Why does this complex carbohydrate "cotton" get so much attention? Cotton's many unique properties have made it useful as a textile fiber for over 5000 years. It is the most imporotant natural textile fiber used in spinning to produce apparel, home furnishings and industrial products—about 40% of all textile fiber consumed in the world. Annually, there is over 25 million metric tons of cotton produced in about 80 countries in the world and it is one of the lead crops to be genetically engineered.

Despite over 150 years of laboratory research on cotton and cellulose, there is much exciting new work to learn about. This book's nationally and internationally recognized contributors have spent their careers gaining in-depth knowledge of cotton fiber chemistry and technology. This includes information on cotton's origin, biosynthesis, production, structural properties/crystal structure of cellulose, morphology, chemical properties/reaction characteristics, degratation, purification, physical properties, classification and characterization (classing), and utilization which is discussed in this book. Some 650 references cited provide a more in-depth treatment of these subjects.

In particular, we acknowledge professor Menachem Lewin for his many important contributions to cotton chemistry and for inspiring this book. Without his strong support it would not exist. Glenn P. Johnson, ARS, SRRC, USDA was invaluable in converting and updating the previous chapter to this book. Fortunately, this book was mostly finished before hurricane Katrina devastated the city of New Orleans and the USDA Southern Regional Research Center where many of the contributors to this book worked.

In conclusion, we have striven to provide information about the chemistry and technology of cotton fiber. We hope this will assist readers in becoming more informed about the unusual carbohydrate called cotton and why it is such an important textile fiber.

Phillip J. Wakelyn
National Cotton Council, Washington, D.C.

Alfred D. French
USDA-ARS-SSRC, New Orleans, LA

Contributors

Noelle R. Bertoniere
Southern Regional Research Center
Agricultural Research Service
U.S. Department of Agriculture
New Orleans, Louisiana

J. Vincent Edwards
Southern Regional Research Center
Agricultural Research Service
U.S. Department of Agriculture
New Orleans, Louisiana

Alfred D. French
Southern Regional Research Center
Agricultural Research Service
U.S. Department of Agriculture
New Orleans, Louisiana

Gary R. Gamble
Cotton Quality Research Station
Agricultural Research Service
U.S. Department of Agriculture
Clemson, South Carolina

Wilton R. Goynes, Jr.
Southern Regional Research Center
Agricultural Research Service
U.S. Department of Agriculture
New Orleans, Louisiana

Lawrance Hunter
Council for Scientific and Industrial
   Research
Port Elizabeth, South Africa

David D. McAlister
Cotton Quality Research Station
Agricultural Research Service
U.S. Department of Agriculture
Clemson, South Carolina

Marie-Alice Rousselle
Southern Regional Research Center
Agricultural Research Service
U.S. Department of Agriculture
New Orleans, Louisiana

Devron P. Thibodeaux
Southern Regional Research Center
Agricultural Research Service
U.S. Department of Agriculture
New Orleans, Louisiana

Barbara A. Triplett
Southern Regional Research Center
Agricultural Research Service
U.S. Department of Agriculture
New Orleans, Louisiana

Phillip J. Wakelyn
National Cotton Council of America
Washington, D.C.

Table of Contents

1 General Description of Cotton

Cotton (Figure 1.1) is the most important natural textile fiber, as well as cellulosic textile fiber, in the world, used to produce apparel, home furnishings, and industrial products. Worldwide about 40% of the fiber consumed in 2004 was cotton [1]. (See also Table 9.1 World Production of Textile Fibers on page 130.) Cotton is grown mostly for fiber but it is also a food crop (cottonseed)—the major end uses for cottonseeds are vegetable oil for human consumption; whole seed, meal, and hulls for animal feed; and linters for batting and chemical cellulose.

Its origin, development, morphology, chemistry, purification, and utilization have been discussed by many authors [2–12]. The chemistry, structure, and reaction characteristics of cellulose, the carbohydrate polymer that forms the fiber, are thoroughly treated in a number of excellent works [8,9,12–19]. This book is intended to provide an overview of the current state of knowledge of the cotton fiber. Much of the information reported here is taken from the references cited at the end of the book, which should be consulted for a more in-depth treatment.

Cotton fibers are seed hairs from plants of the order Malvales, family Malvaceae, tribe Gossypieae, and genus Gossypium [2–5,10,11]. Botanically, there are four principal domesticated species of cotton of commercial importance: *hirsutum, barbadense, aboreum, and herbaceum*. Thirty-three species are currently recognized; however, all but these four are wild shrubs of no commercial value. Each one of the commercially important species contains many different varieties developed through breeding programs to produce cottons with continually improving properties (e.g., faster maturing, increased yields, and improved insect and disease resistance) and fibers with greater length, strength, and uniformity.

Gossypium hirsutum, a tetraploid, has been developed in the United States from cotton native to Mexico and Central America and includes all of the many commercial varieties of American Upland cotton. Upland cottons now provide over 90% of the current world production of raw cotton fiber. The lengths, or staple lengths, of the Upland cotton fiber vary from about $\frac{7}{8}$ to $1\frac{1}{2}$ in. (22–36 mm), and the micronaire value (an indicator of fiber fineness and maturity but not necessarily a reliable measure of either; see Chapter 8) ranges from 3.8 to 5.0. If grown in the United States, *G. hirsutum* lint fibers are 26–30 mm (1 to 1–3/16 in.) long [20]. Fiber from *G. hirsutum* is widely used in apparel, home furnishings, and industrial products.

Gossypium barbadense, a tetraploid, is of early South American origin and provides the longest staple lengths. The fiber is long and fine with a staple length usually greater than $1\frac{3}{8}$ in. (35 mm) and a micronaire value of below 4.0. If grown in the United States., *G. barbadense* lint fibers are usually 33–36 mm ($1\frac{5}{6}$ to $1\frac{1}{2}$ in.) long [20]. Commonly known as extra-long-staple (ELS), it supplies about 8% of the current world production of cotton fiber. This group includes the commercial varieties of Egyptian, American–Egyptian, and Sea Island cottons. Egypt and Sudan are the primary producers of ELS cottons in the world today. Pima, which is also ELS cotton, is a complex cross of Egyptian and American Upland strains and is grown in the western United States (mainly California with some in Arizona, southwestern Texas, and New Mexico), as well as in South America. Pima has many of the characteristics of the

FIGURE 1.1 Mature cotton in the field ready for harvesting. (Courtesy of the National Cotton Council of America, Memphis, TN.)

better Egyptian cottons. This fiber from *G. barbadense* is used for the production of high-quality apparel, luxury fabrics, specialty yarns for lace and knitted goods, and sewing thread.

The other commercial species—*Gossypium aboreum* and *Gossypium herbaceum*, both diploids—are known collectively as "Desi" cottons, and are the Asiatic or Old World short staple cottons. These rough cottons are the shortest staple cottons cultivated (ranging from $\frac{3}{8}$ to $\frac{3}{4}$ in. (9.5–19 mm)) and are coarse (micronaire value greater than 6.0) compared with the American Upland varieties. Both are of minor commercial importance worldwide but are still grown commercially in Pakistan and India. *G. aboreum* is also grown commercially in Burma, Bangladesh, Thailand, and Vietnam [3,10].

Varietal development programs were once confined to classical methods of breeding that rely on crossing parents within species. Currently, in addition to the conventional breeding methods, research is underway on hybrids to produce new varieties, and modern biotechnology (recombinant DNA technology) to produce biotech or transgenic cottons [21,22], which enhance production flexibility. Biotech cottons deliver high-tech options to farmers and consumers without compromising environmental quality. Since the introduction of *Bacillus thuringiensis* (Bt) biotech cotton in 1996, cotton has been one of the lead crops to be genetically engineered, and biotech cotton has been one of the most rapidly adopted technologies ever. The current varieties of commercial importance address crop management or agronomic traits that assist with pest management (insect resistant) or weed control (herbicide tolerance). Nine countries, representing over 59% of world cotton area, allow biotech cotton to be grown: Argentina, Australia, China (Mainland), Colombia, India, Indonesia, Mexico, South Africa, and United States. Other countries (e.g., Pakistan, Brazil, Burkina Faso, Egypt, and the Philippines) are considering approving the cultivation of biotech cotton. In 2003–2004, about 21% of the world's cotton acreage and about 30% of the cotton produced in the world was biotech cotton; in 2004–2005, about 35% of world's cotton production was biotech cotton; and within five years, world biotech cotton production could be close to 50%.

The initial biotech efforts have been centered on insect resistance and herbicide tolerance. Insect resistance has been conferred through the incorporation of genes from Bt that produce Bt δ-endotoxin, a naturally occurring insect poison for bollworms and budworms. The reduction in the use of insecticides minimizes the adverse effects on nontarget species and beneficial insects. Herbicide tolerance enables reduced use of herbicides and encourages use of safer, less persistent materials to control a wide spectrum of weeds that reduce yield and lint quality of cotton. Transgenic herbicide tolerance moves cottonweed management away from protective, presumptive treatments toward responsive, as-needed treatments. While insect resistance and herbicide tolerance are the only traits currently available in biotech cottons, a broad range of other traits are under development using modern biotechnology. These may impact the agronomic performance, stress tolerance, fiber quality, and yield potential directly. However, in 2005 few of these traits are close to commercialization. As soon as new developments in bioengineered cotton for insect resistance, herbicide tolerance, stress tolerance, yield potential, improved fiber quality, etc., are available, they will be incorporated into conventional cotton varieties.

The cottons of commerce are almost all white (creamy yellow to bright white). In recent years, there has been a renewed interest in naturally pigmented and colored cottons, which have existed for over 5000 years [23,24]. These cotton varieties are spontaneous mutants of plants that normally produce white fiber. The availability of inexpensive dyes and the need for higher-output cotton production worldwide caused the naturally colored cottons to almost disappear about 50 years ago. Yields were low and the fiber was essentially too short and weak to be machine spun. Breeding research over the last 15 years reportedly has led to improvement in yields, fiber quality, fiber length and strength, and color intensity and variation [25]. Naturally colored cottons are a very small niche market. The cottons available today are usually shorter, weaker, and finer than regular Upland cottons, but they can be spun successfully into ring and rotor yarns for many applications [26]. They can be blended with normal white cottons or blended among themselves. For a limited number of colors, the use of dyes and other chemicals can be completely omitted in textile finishing, possibly generating some savings, which can compensate for the higher raw material price. The color of the manufactured goods can intensify with washing (up to 5 to 10 washings), and colors vary somewhat from batch to batch [26]. Naturally colored cottons are presently grown in China, Peru, and Israel. The amount available in 2005 is very small, perhaps 10,000 U.S. bale equivalents (about 2270 metric tons). Shades of brown and greens are the main colors that are available. Other colors (mauve, mocha, red) are available in Peru in a very limited supply and some others are under research. The color for brown and red-brown cottons appears to be in vacuolar tannin material bodies in the lumen (Figure 1.2a). The different colors of brown and red-brown are mostly due to catechin–tannins and protein–tannin polymers [27]. The green color in cottons (Figure 1.2b) is due to a lipid biopolymer (suberin) sandwiched between the lamellae of cellulose microfibrils in the secondary wall [27–29]. The brown fibers (and white lint) do not contain suberin. Green cotton fibers are characterized by high wax content (14–17% of the dry weight) whereas white and brown fibers contain about 0.4–1.0% wax [27].

Cotton grown without the use of any synthetically compounded chemicals (i.e., pesticides, fertilizers, defoliants, etc.) is considered as "organic" cotton [30–34]. It is produced under a system of production and processing that seeks to maintain soil fertility and the ecological environment of the crop. To be sold as organic it must be certified. Certified organic cotton was introduced in 1989–1990 and over 20 countries have tried to produce organic cotton. Since 2001, Turkey has been the largest producer of organic cotton. There are small projects in Mali, Kyrgyzstan, and some other developing countries [35]. In 2001, about 14 countries in the world produced about 27,000 U.S. bale equivalents (5700 metric tons) of organic cotton, with Turkey, the United States, and India accounting for about 75% of production [36]. In 2003, in the United States, 4628 U.S. bale equivalents of organic cotton were produced; in 2004, 4674 acres were planted and about 5000 U.S. bale equivalents were produced [37]. In 2005, the world's production of organic cotton was about 100,000 U.S. bale equivalents (about 25,000 metric tons), which is less than 0.12% of world's total cotton production [38].

Unlike synthetic fibers, which are spun from synthetic or regenerated polymers in factories, cotton fiber is a natural agricultural product. The United States and some other countries use the newest and latest tested technology to produce the cotton crop [8,9]. Cotton

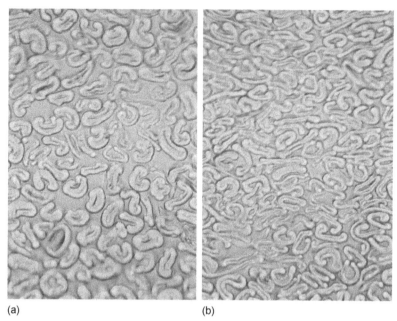

(a)   (b)

FIGURE 1.2 Fiber bundle cross-sections obtained with transmitted light microscopy (a) for natural brown cotton where the presence of material bodies are visible within the lumens of some fibers and (b) for natural green cotton where the fibers are quite immature and are chacterized by the presence of suberin in the fiber walls and not material bodies in the lumen.

production is scattered among hundreds of thousands of small, independent farms (about 20,000–25,000 in the United States), whereas the synthetic fiber industry is concentrated in a few corporations that have production plants worldwide. Cotton is grown in about 80 countries in the world; in 2004–2005, 59 countries cultivate at least 5000 ha (12,350 acres (1 ha = 2.47 acres)) [39]. In many developing countries, cotton farms range in size from less than 1 to about 10 ha. In 2004, developing countries accounted for about 75% of the world's production, and China, the United States, India, Pakistan, Uzbekistan, Turkey, and Brazil account for over 81% of the world's cotton production [40]. Cotton production in the United States is usually about 18–20 million U.S. bale equivalents (8.6–9.6 billion pounds (3.9–4.4 million metric tons)) annually and worldwide production is about 90–105 million bales (45.0–52.5 billion pounds (20.4–23.8 million metric tons)) annually. Raw cotton is exported from about 57 countries and cotton textiles from about 65 countries.

Cotton production, harvesting, and ginning are described in more detail in other sources [2–5,8,9,41,42]. The cotton plant is a tree or a shrub that grows naturally as a perennial, but for commercial purposes it is grown as an annual crop. Botanically, cotton bolls are fruits. Cotton is a warm-weather plant, cultivated in both hemispheres, mostly in North and South America, Asia, Africa, and India (in tropical latitudes). Mostly it is cultivated in the Northern Hemisphere. It is primarily grown between 37°N and 32°S but can be grown as far north as 43°N latitude in Central Asia and 45°N in mainland China. Planting time for cotton varies with locality, i.e., from February to June in the Northern Hemisphere [2–5,8–11]. The time of planting in the Northern Hemisphere is the time of harvest in the Southern Hemisphere. Seedlings emerge from the soil within a week or two after planting, 5–6 weeks later flower buds or squares form, and white (Upland cotton) or creamy yellow (Pima cotton) to dark-yellow blossoms appear in another 3–4 weeks. The time interval from bloom to open boll is about 40–80 days. The open boll lets air in to dry the white, clean, fiber, and fluff it for the harvest (Figure 1.3).

FIGURE 1.3 Open cotton boll. (Courtesy of the National Cotton Council of America, Memphis, TN.)

Each cotton fiber is a single, elongated, complete cell that develops in the surface layer of cells of the cottonseed. The mature cotton fiber is actually a dead, hollow, dried cell wall [4,5,43]. In the dried out fiber, the tubular structure is collapsed and twisted, giving cotton fiber convolutions, which differentiate cotton fibers from all other forms of seed hairs and are partially responsible for many of the unique characteristics of cotton. The biosynthesis and morphology of the cotton fiber are discussed in more detail later (see Chapter 2).

The relatively long fiber lengths (about 1 in. (25.4 mm) or longer) on the cottonseed relate to the fiber that is used by the textile industry. This raw cotton fiber, which can be spun into textile yarns, is called lint. However, another type of fiber, *linters* or *fuzz fibers*, which are very short, is also produced on the seed along with the lint [44]. The distribution of the lint and fuzz fibers over the seed surface is neither uniform nor random. The base of the seed mostly produces lint fibers, whereas cells near the tip of the seed mostly produce fuzz fibers.

The long lint fibers are removed at the cotton gin, but the short linter fibers are still attached to the seed after ginning for *G. hirsutum* (Upland cotton); *G. barbadense* seed does not contain linters. The linter fibers are removed at the cottonseed oil mill by the delintering process prior to the oil extraction process, producing fibers of different lengths. In general, ginned cottonseed is composed of about 8% linters. Linters can be distinguished from lint by several characteristics:

1. Length (commercial lint averages 1 in. (25.4 mm) and first-cut linters average about 0.5 in. (12–15 mm)
2. Pigmentation (linters are often light brown or highly colored)
3. Strength of adherence to seed (linters are more tightly held to the seed)
4. Chemical and physical properties

Linters are also much coarser than lint fibers; linters have a coarse, stiff form without the flexibility or convolutions of lint fibers; and the tips of linter fibers are somewhat tapered but not to the same extent as those of lint fibers. Linter fibers have hardly any lumen. The diameter of the linter fiber is usually about twice the diameter of the lint fiber. In lint fibers, the growth of the primary cell wall starts on the day of flowering and within 20 days they exhibit a 3000-fold increase in length, whereas fuzz fibers (linters) initiate elongation about 4 days after flowering. The secondary cell wall, formation of which begins about 20 days after flowering, is different in lint and in fuzz fibers; for example, the secondary cell wall at the base of the fiber is much thicker in linters than in lint. Although not suitable for textile processing, the linters are used in other applications, for example, second-cut linters (the shortest linter fiber) are used as a chemical feedstock for manufacturing plastics and rayons, whereas first-cut linters (the longer fiber) can be used for batting and padding in bedding, upholstered furniture, automotive applications, and paper (most currencies are made of cotton paper).

Harvesting time for cotton varies with locality (Table 1.1).

TABLE 1.1
Typical Weight and Densities of Cotton Bales in Various Countries and Harvest Date

		Bale Weight				Density	
		Avg.		Range		kg/m³	lb/ft³
Country	Harvest	kg	lb	kg	lb		
North and Central America							
U.S.	July–Jan.	225	495	217–230	477–506	427.1	26.7
Mexico	June–Jan.	227	499	180–240	396–528	380.1	23.8
Guatemala	Nov.–Mar.	227	499	220–236	484–519	251.4	15.7
South America							
Brazil	Aug.–Jan. or Feb.–May	140	308	110–180	242–396	243.1	15.2
Argentina	Feb.–June	220	484	195–250	429–550	346–429	21.6–26.8
Paraguay	Feb.–June	198	436	160–225	352–495	420.9	26.3
Columbia	July–Sept. or Dec.–Mar.	225	495	220–240	484–528	516.2	32.3
Peru	Feb.–Oct.	240	528				
Venezuela	Feb.–May						
Europe							
Greece	Sept.–Nov.	234	516				
Spain	Sept.–Nov.	225	495	200–250	440–550	396.8	24.8
Asia–Oceania							
Uzbekistan	Sept.–Nov.	200	440	190–220	418–484	490.2	30.6
China Bale I	Sept.–Nov.	85	187	80–90	176–198	442.7	27.7
Bale II	Sept.–Nov.	200	440	190–210	418–462	439.8	27.5
India	July–Jan. or Dec.–May	170	375	165–175	363–385	379.8	23.7
Pakistan	Sept.–Feb.	170	375	160–185	352–407	547.5	34.2
Turkey	Sept.–Dec.	217	478	190–245	418–539	342.3	21.4
Australia	Apr.–June	227	499	150–240	330–528	449.6	28.1
						327.6	20.4
Iran	Oct.–Dec.	200	440	170–240	375–528	226.8	14.1
Syria	Sept.–Nov.	205	451	192–212	422–466	425–434	26.6–27.1
Africa							
Egypt	Sept.–Oct.	327	720	290–345	638–759	541.6	33.8
Sudan	Jan.–Apr. or Sept.–May	188	414	185–191	407–420	361–397	22.6–24.8
South Africa	Apr.–May	206	453	170–240	375–528	437.4	27.3
Ivory Coast	Oct.–Jan.	215	474				
Tanzania	May–July	183	403				
Nigeria	Dec.–Feb.	185	407				

Source: Bale weights—From Munro, J.M., *Cotton*, 2nd ed., Longman Scientific and Technical, Essex, England and John Wiley & Sons, New York, 1987, p. 333; Bale survey—From International Cotton Advisory Committee, Oct. 1995; and Harvest date—From Volkart Brothers Holding, Ltd., Switzerland, 1991.

(a)   (b)

FIGURE 1.4   (a) Mechanical harvesting by means of a cotton picker. (b) Mechanical harvesting by means of a cotton stripper. (Courtesy of the National Cotton Council of America, Memphis, TN.)

The operations of harvesting [11] and ginning the fiber [41], as well as cultural practices during the growing season, are very important to the quality of the cotton fiber [5,11]. Harvesting is one of the final and most important steps in the production of a cotton crop [5,9,11,12], as the crop must be harvested before the inclement weather can damage the quality and reduce the yield. Because of economic factors, virtually the entire crop (>99%) in the United States and Australia is harvested mechanically (Figure 1.4). In rest of the world (~75%) hand harvesting of cotton, one boll at a time, is still quite prevalent, particularly in the less developed countries and in countries where the labor is cheaper [45].

Mechanically harvested cotton, either with cotton picker machines (cotton burr remains attached to the stalk) or with stripper machines (cotton burr is removed along with the seed cotton), can contain more trash and other irregularities than hand-harvested cotton. However, according to "Cotton contamination surveys" by the International Textile Manufactures Federation (ITMF), the most contaminated cottons originate from some of the countries where cotton is hand-picked, whereas some of the cleanest can be sourced in the USA where cotton is machine harvested [34]. Most of the mechanically harvested cotton is harvested with cotton pickers (~75% in the United States and all in Australia).

After harvesting, the seed cotton (consisting of cotton fiber attached to cottonseed and plant foreign matter), a raw perishable commodity, is transported to the ginning plant in trailers or modules, or is stored in the field in modules. In the United States, module storage is used for almost the entire crop. Field storage in modules maximizes efficiency at the gin.

Ginning is the separation of the fibers from seed and plant foreign matter [41,42]. Cotton essentially has no commercial value or use until the fiber is separated from the cottonseed and the foreign matter at the gin. Ginning operations, which are considered a part of the harvest, are normally considered to include conditioning (to adjust moisture content), seed–fiber separation, cleaning (to remove plant trash), and packaging (Figure 1.5). Upland cottons are ginned on saw gins (Figure 1.6), whereas roller gins are used for ELS cottons.

Raw cotton in its marketed form consists of masses of fibers in densely packed bales (>22 lb/ft^3 (352 kg/m^3), Figure 1.7) [42].

The bales into which cotton is packed are of varying dimensions, volumes, densities, and weights (see Table 1.1) and are mainly covered with woven polypropylene, polyethylene film, burlap, or cotton fabrics (Figure 1.8).

In the United States, bales weigh on an average 490 lb (222 kg). A single pound of cotton may contain 100 million or more individual fibers (about 50 to 55 billion fibers in a 480-lb bale).

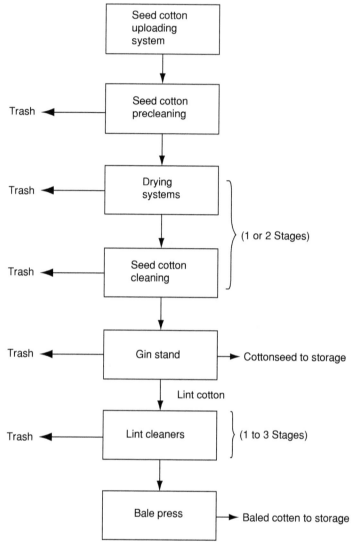

FIGURE 1.5 Simplified flow chart of ginning systems. (Courtesy of the National Cotton Council of America, Memphis, TN.)

In the rest of the world, bales usually weigh 375–515 lb (170–233 kg), depending on the country where they are produced (Table 1.1) [19,41]. The *International Organization for Standardization* (ISO) specifies (ISO 8115 [46]) that bale dimensions and densities should be: length 1400 mm (55 in.), width 530 mm (21 in.), height 700–900 mm (27.5–35.4 in.), and density of 360–450 kg/m^3 (22.5–28.1 lb/ft^3). Typical bale dimensions for the U.S. universal density (UD) bales are 1400 mm (55 in.) length × 533 mm (21 in.) width × 736 mm (29 in.) height, resulting in a bale density of approximately 448 kg/m^3 (28 lb/ft^3).

At the gin, baled cotton is sampled so that grade and other quality parameters can be determined and the cotton is classed at the U.S. Department of Agriculture (USDA) classing offices. Classification is a way of measuring the fiber quality and physical attributes of this natural product that affect the manufacturing efficiency and quality of the finished product (see Chapter 8). Cotton bales are usually stored in warehouses prior to going to the textile mill [42].

FIGURE 1.6 Cotton gin stand, saw-ginning process. (Courtesy of the National Cotton Council of America, Memphis, TN.)

Cotton is merchandized and shipped by a cotton merchant prior to arriving at the textile mill, where it is manufactured into products for the ultimate consumer. The marketing of cotton is a complex operation that includes transactions involving the buying, selling, or reselling from the time the cotton is ginned until it reaches the textile mill. In the United States, after the cotton is ginned and baled, growers usually sell their cotton to a merchant or store it in

FIGURE 1.7 Packaged bales (universal density) ready to be shipped to the textile mill. (Courtesy of the National Cotton Council of America, Memphis, TN.)

FIGURE 1.8 A newly bound bale comes off the press at conclusion of ginning process. (Courtesy of the National Cotton Council of America, Memphis, TN.)

a government-approved warehouse and borrow money against it. In some other countries, the cotton is sold as seed cotton and the buyer has it ginned and then sells the lint cotton.

When cotton is grown and processed in a responsible manner, it does not have adverse effects on the environment, the workplace, and the consumer [47] (see Chapter 10).

2 Biosynthesis of Cotton

The cotton fibers used in textile commerce are the dried cell walls of formerly living cells. Botanically, cotton fibers are trichomes or seed coat hairs that differentiate from epidermal cells of the developing cottonseed. The cotton flower blooms only for one day and quickly becomes senescent thereafter. On the day of full bloom, or anthesis, the flower petals are pure white in most *G. hirsutum* varieties. By the day after anthesis, the petals turn bright pink in color and, usually by the second day after anthesis, the petals fall off the developing carpel (boll). The day of anthesis serves as a reference point for all subsequent events in the seed and fiber development.

Typically, there are 20–30 ovules in a boll containing three to five segmented compartments (locules). The ovules are attached to the plant via a connection called the funiculus. Fertilization of the ovules is essential for subsequent development of the seed and fiber. When there is a failure in the fertilization of a seed, development is aborted and the resulting mote may cause problems during later fiber and fabric processing steps. It is possible to remove unfertilized ovules from the boll and mimic the conditions for seed and fiber growth in tissue culture [48]. Two phytohormones, namely auxin and gibberellic acid, are required in ovule culture to stimulate seed and fiber development.

On, or slightly before the day of anthesis, the morphological events marking the initiation of fiber development are evident. Lint fiber development initiates first at the more rounded end (chalaza) of the seed and proceeds around the seed surface to the micropyle. The fiber cells first assume a fairly rounded or bulbous appearance and are visible above the formerly smooth ovule epidermal surface. Approximately one out of every four epidermal cells begins this cellular differentiation process [49]. The physiological and biochemical factors that regulate which epidermal cells will become fiber cells are unknown at present. Approximately 6–7 days later, a second type of fiber cells called fuzz fibers (linters) begin growing. Fuzz fibers are distinguished from lint fibers by their larger perimeter, shorter length, and final chemical composition [50].

In the next phase of lint fiber development, the cells expand longitudinally, reaching their final lengths in 21–35 days postanthesis (DPA). The ultimate length of fiber cells is controlled genetically by different cotton genotypes having fiber lengths of nearly 1 to 6 cm. Abnormal environmental conditions during cell elongation can alter the ability of fiber cells to reach their full potential length.

During the elongation phase of fiber development, the cell is delimited by a primary cell wall and covered by a waxy layer or cuticle (Figure 2.1).

Recently, genotypic variation in mature cotton fiber surface components, presumably from the cuticle, has been reported [51]. In the cytoplasm, smaller vacuoles coalesce into one large central vacuole leaving only a thin ribbon of cytoplasm between the vacuole membrane and the cellular membrane (plasmalemma). Organelles are distributed in the cytoplasm in a manner that is consistent with a model for cell expansion occurring by intercalation rather than tip growth [52,53]. The random orientation of cell wall polymers in the primary cell wall

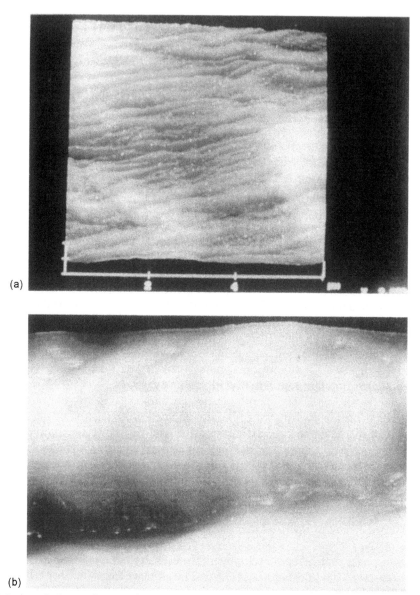

(a)

(b)

FIGURE 2.1 Atomic force micrographs of the cotton fiber cuticular layer on the surface of unprocessed fiber. (a) Cuticular material is deposited in undulating waves (16-μm scan). (b) Irregular deposition of cuticular material at higher magnification (550-nm scan). (Courtesy of T. Pesacreta, University of Southwestern Louisiana, and L. Groom, USDA-Forest Service.)

leads to an interwoven network of carbohydrate and protein macromolecules. New primary cell wall material must be coordinately deposited into the exocellular matrix, while at the same time the integrity of the primary cell wall must not be broken. Cotton fiber primary cell wall structure changes during development as demonstrated by changes in sugar composition and molecular mass distribution of the noncellulosic polysaccharides, including xyloglucan [54,55]. The biochemistry of primary cell wall expansion is an active area of investigation in plant biology [56,57]. As in other plant cell types, the biosynthesis and modification of most of the primary cell wall polymers are believed to occur in Golgi bodies with subsequent transport to the plasmalemma in small vesicles. In contrast, cellulose microfibrils of the secondary wall are synthesized by multisubunit enzyme complexes associated with the cell membrane.

There has been little direct characterization of enzymes that are responsible for the synthesis of cotton fiber cell wall components; however, application of molecular genetic techniques in recent years has greatly increased our knowledge about the structure of genes coding for these enzymes. Several large-scale efforts to characterize expressed sequence tags (ESTs) from elongating cotton fiber have resulted in an enormous increase in gene sequence information for the genus *Gossypium*. ESTs are partial or incomplete DNA sequences corresponding to the genes that are transcribed (active) in a particular tissue or stage of development. In public databases, such as GenBank at the National Center for Biotechnology Information (http://www.ncbi.nlm.nih.gov), comparisons of cotton fiber ESTs can be made with genes from other organisms. These searches predict the most likely match based on nucleotide sequence similarity or amino acid sequence similarity from deduced proteins. Estimates are that >2500 genes or over 14% of the currently available EST sequences for *G. arboreum* are related to primary cell wall metabolism and another 12% of the EST sequences are related to secondary cell wall biogenesis [58].

At approximately 15 to 19 DPA, lint fiber cells begin producing a secondary cell wall composed of highly crystalline, nearly pure cellulose, a β-(1→4)-D-glucan. The nature of the signal responsible for triggering secondary cell wall biosynthesis is unknown. Cellulose synthesis continues for 30–45 days until the cell wall is 2–6 μm thick. By the time the fiber is mature, over 90% of the dry weight of the fiber is cellulose. Callose, a β-(1→3)-D-glucan polymer is also transiently synthesized during the transition phase of development [59–61] and is deposited between the thickening secondary wall and the plasmalemma [62]. The function of callose in cotton fiber remains an enigma; however, recent evidence suggests that the cotton gene coding for callose synthase is not identical to the gene for cellulose synthase as had been proposed for many years [63].

In the secondary cell wall, cellulose microfibrils are deposited in a very ordered fashion. Concentric lamellae of cellulose microfibrils are subsequently deposited in a helical manner with a gradual increase in pitch. Periodically, the gyre of the helical microfibril orientation changes. These changes in the gyre are called reversals and have been shown to occur at or near where fiber breakage is likely [64,65]. Cellulose microfibril orientation angle is an important factor in fiber strength [66]. The orientation of cellulose microfibrils in the cotton fiber secondary cell wall is regulated by the orientation of structures in the cytoplasm collectively called the cytoskeleton [67]. Principal components of the cytoskeleton are the filamentous structures known as microtubules and microfilaments. Microtubules are composed of two proteins, α- and β-tubulins, and microfilaments are composed of the protein, actin. Regulation of the formation, organization, and disassembly of the cytoskeleton is believed to be controlled by other, less abundant proteins associated with the major cytoskeletal proteins. The characterization of these cytoskeletal components, their regulatory molecules, and how they might dictate cellulose microfibril organization in cotton fiber is an ongoing research effort [68–72].

Although sucrose is the primary transport sugar in the developing boll, the primary substrate for the enzymatic synthesis of cellulose is UDP–glucose [73]. Particulate forms of the enzyme, sucrose synthase [SuSy; EC 2.4.1.13], are believed to synthesize most of the UDP–glucose in the reversible reaction, sucrose + UDP ↔ UDP–glucose + fructose [74]. The case for SuSy involvement in cellulose biosynthesis is enhanced by the finding that SuSy localizes in the same regions of the cell where cellulose biosynthesis is predicted to occur [75]. β-sitosterol-D-glucoside, a lipid-linked sugar, has been proposed to function as a primer for cellulose biosynthesis [76,77]; however, *Arabidopsis* mutants that are genetically impaired in sitosterol synthesis can still produce cellulose [12].

Structures associated with cellulose biosynthesis in higher plants were first characterized [78] by freeze-fracture microscopy. Arranged in the plasma membrane, these hexagonal structures, called rosettes, are assembled in the Golgi apparatus [79]. The six particles making up a rosette are composed of four to six multisubunit complexes, each one of which synthesizes a molecule of cellulose. Due to numerous difficulties in purifying enzyme complexes from higher plants capable of synthesizing cellulose [80–82], attention has focused away from the enzymology of cellulose biosynthesis and diverted toward characterization of genes encoding the enzymes. A breakthrough was achieved when ESTs from cotton fiber were shown to be somewhat similar to short regions in the DNA sequence of a cellulose synthase from the bacterium, *Acetobacter xylinum* [83]. The regions in the bacterial gene that corresponded to two cotton ESTs share sequence similarity with the binding site and catalytic site of the glycosyltransferase family of genes, of which cellulose synthase is a member. The genes from cotton were named *CesA1* and *CesA2*, and were the first clones ever characterized for a higher plant cellulose synthase (*CesA*; EC 2.4.1.12). An antibody to cotton *CesA1* was used to demonstrate directly that the catalytic subunit for cellulose biosynthesis colocalizes with rosettes [84].

Since discovery of the catalytic subunit gene in cotton, *CesA* genes have been cloned from many different plants, most notably from the model organism, *Arabidopsis thaliana*, whose genome has been fully sequenced. Genomic analysis of *Arabidopsis* indicates that there are at least 10 genes in the true *CesA* subfamily and a multitude of other related genes in six subfamilies of cellulose synthase-like (*Csl*) [85]. Multiple forms of the *Arabidopsis CesA* genes must be expressed in the same cell [86]. Current evidence suggests that, in *Arabidopsis,* three distinct *CesA* genes code for the enzyme that produces cellulose for the primary wall [87,88] and three other genes code for the enzymes that produce cellulose in the secondary cell wall [89,90].

The first two higher plant cellulose synthase genes cloned, *GhCesA1* and *GhCesA2,* are transcribed during the secondary wall thickening phase of cotton fiber development [25]. A putative third cellulose synthase gene that is expressed during the same developmental period has recently been cloned and characterized [91]. Measurements of relative transcript abundance for the cotton *CesA* genes indicate that some members of this gene family are transcribed during the cell elongation stage of fiber development and other *CesA* genes are transcribed during secondary cell wall formation [55]. Cotton is an allotetraploid with part of the genome contributed by one diploid progenitor (A subgenome), and the remainder of the genome contributed by another diploid progenitor (D subgenome). Interestingly, cotton *CesA* genes coded by both subgenomes appear to be transcribed [55,92].

The molecular weight of primary cell wall cellulose is less than the molecular weight of cellulose in the secondary cell wall [93,94]. Based on the temporal regulation of *CesA* gene expression, this difference in cellulose molecular weight may be due to the two different types of cellulose synthase complexes formed during the two phases of fiber development. Difficulty in reconstituting enzyme activity from cloned plant genes has prevented a direct resolution of this issue. Another complication is that in addition to *CesA*, several other genes are important for cellulose production as demonstrated by genetic approaches in *Arabidopsis*. These additional genes include those coding for a β-(1→4)-glucanase gene (*KOR*), a serine-rich protein (*KOB1*), three enzymes involved in N-linked glycan formation (*KNF, rsw3, CYT1*), three sterol biosynthesis genes (*FK, HYD1, SMT1/CPH*), a GPI-anchored COBRA protein (*COBL1*), and a chitinase-like gene (*CTL1*). How these gene products are involved in cellulose biosynthesis remains unknown at present. Genomic approaches to identify orthologs of these genes from cotton are ongoing [95].

Cotton bolls dehisce at maturity, leaving the fibers fully exposed to air and sunlight. The water content of the fiber decreases rapidly, the cytoplasm dries against the inner surface of the wall, and a large lumen is left where the central vacuole was once located. The formerly tube-shaped cell collapses and assumes a twisted ribbon conformation with a kidney-like cross-sectional pattern. These twists, or convolutions, permit the spinning of fiber cells into yarns.

3 Chemical Composition of Cotton

Raw cotton fiber, after ginning and mechanical cleaning, is approximately 95% cellulose (Table 3.1) [96–99]. The structure of cotton cellulose, a linear polymer of β-D-glucopyranose, is discussed in Chapter 5 and chemical properties in Chapter 6.

The noncellulosic constituents of the fiber are located principally in the cuticle, in the primary cell wall, and in the lumen. Cotton fibers that have a high ratio of surface area to linear density generally exhibit a relatively higher noncellulosic content. The noncellulosic constituents include proteins, amino acids, other nitrogen-containing compounds, wax, pectic substances, organic acids, sugars, inorganic salts, and a very small amount of pigments. Variations in these constituents arise due to differences in fiber maturity, variety of cotton, and environmental conditions (soil, climate, farming practices, etc.). After treatments to remove the naturally occurring noncellulosic materials, the cellulose content of the fiber is over 99%.

The noncellulosic materials can be removed by selective solvents. The wax constituent can be removed selectively with nonpolar solvents, such as hexane and chloroform, or nonselectively by heating in a 1% sodium hydroxide solution. Hot nonpolar solvents and other water-immiscible organic solvents remove wax but no other impurity, hot ethanol removes wax, sugars, and some ash-producing material but no protein or pectin, and water removes inorganic salts (metals), sugars, amino acids and low-molecular-weight peptides, and proteins. Most of the nonpolymeric constituents including sugars, amino acids, organic acids, and inorganic salts may be removed with water. The remaining pectins and high-molecular-weight proteins are removed by heating in a 1% sodium hydroxide solution or by appropriate enzyme treatments. All of the noncellulosic materials are removed almost completely by boiling the fiber in hot, dilute, aqueous sodium hydroxide (scouring or kier boiling), then washing thoroughly with water.

Of all the noncellulosic constituents, the nitrogen-containing compounds constitute the largest percentage when expressed as percent protein (1.1–1.9%) [6,7,99,100]. The nitrogen content of scoured fiber is about 0.035% (about 0.22% protein). Most of the nitrogenous material occurs in the lumen of the fiber, most likely as protoplasmic residue [101], although a small portion is also extracted from the primary wall [102]. The nitrogen-containing compounds located in the lumen may be removed using water, while those located in the primary cell wall are removed by heating in a 1% sodium hydroxide solution (a mild alkali scour such as that used to prepare cotton fabrics for dyeing and finishing). Cotton fiber and its primary wall both contain proteins and peptides, free amino acids, and most likely nonprotein nitrogen [102,103]. The free amino acids that have been detected are glutamic acid, aspartic acid, valine, serine, and threonine [103].

Cotton wax (about 0.4 to 1.0% of fiber dry weight) comprises the cuticle on the outer surface of the fiber. The quantity of wax increases with the surface area of the cotton, and the

TABLE 3.1
Composition of Typical Cotton Fibers

| | Composition (% Dry Weight) | |
Constituent	Typical %	Range %
Cellulose	95.0	88.0–96.0
Protein (% N × 6.25)[a]	1.3	1.1–1.9
Pectic substances	0.9	0.7–1.2
Ash	1.2	0.7–1.6
Wax	0.6	0.4–1.0
Total sugars	0.3	0.1–1.0
Organic acids	0.8	0.5–1.0
Pigment	trace	–
Others	1.4	–

[a]Standard method of estimating percent protein from nitrogen content (% N).

finer cottons tend to have a larger percentage of wax. The wax is a mixture of high molecular weight, primarily long-chain saturated fatty acids and alcohols (with even numbers of carbon atoms, C_{28} to C_{34}), resins, saturated and unsaturated hydrocarbons, sterols, and sterol glucosides [104,105], including montanyl triacontanoate (10–15%), montanol (25%), 1-triacontanol (18%), and β-sitosterol (10%). A mutant variety (*G. hirsutum L.*) of cotton with natural green color related to a suberin-like wax biopolymer is located between cellulose lamellae in the secondary wall [28], with a 14–17% wax composition relative to fiber weight. This waxy polymer is composed predominantly of ω-hydroxydocosanoic acid with which glycerol and the phenolic compound, caffeic acid, are associated [29,106]. Naturally colored brown-lint cottons and other colored cottons seem to have normal wax contents. Other phenolic compounds may be responsible for color formation in the other colored fiber varieties, but the identity of these compounds is not known at present.

The natural wax content serves as a protective barrier both to water penetration and to microbial degradation of the underlying polysaccharides. The wax serves as a lubricant that is essential for proper spinning of cotton fiber into yarn. Once the yarn is spun, however, the wax does reduce the tensile strength of the yarn as well as hinders dyeing and finishing of the fiber. Wax is detrimental in the chemical processing and finishing of the cotton yarns and fabrics because it interferes with wetting of the fiber and penetration of the reagents. If the wax layer is disrupted, for example, by the microbiological action due to weathering, it can affect the sizing operation. The sizes will penetrate too far into the fiber and the fiber will pick up too high a level of size. The cotton wax, therefore, is removed (saponified) by a mild scouring treatment using sodium hydroxide solution in the normal preparation of cotton yarns and fabrics for dyeing and finishing. After the wax is removed from yarn in the scouring operation, the fullest development of strength is attained in the end product.

The content and texture of the wax are important in predicting the processing potential of the cotton. Mild scouring treatments of cotton, like those acceptable to qualify cotton as washed cotton under the U.S. cotton dust standard, sometimes can alter the wax surface and adversely affect textile processing without lowering the amount of wax on the fiber [107].

Underlying the waxy cuticle is the primary cell wall, which is composed of two distinct layers [108]. The outermost layer is comprised primarily of pectin substances (usually designated as pectin) in the form of free pectic acid (linear polymer of (1→4)-D-galacturonic acid) and its insoluble calcium, magnesium, and iron salts, and constitutes approximately 0.7–1.2% of the dry fiber weight. The innermost layer is comprised of hemicelluloses, primarily in the form of xyloglucan, and cellulose. Pectin is removed either by the same scouring process that removes wax as part of preparation for dyeing and finishing, or by a combination of pectinase enzymes. The method of pectin analysis has much to do with the percent pectin reported. Removal of pectin does not significantly alter the tensile strength of the fiber and has little effect on the yarn and the fabric properties.

Soluble sugars (about 0.1 to 1.0% of fiber dry weight) found on cotton originate from two sources: metabolic residues (plant sugars) located within the dried lumen and the outer fiber surface and insect sugars (insect "honeydew" excretion) found on the outer surface of the fiber [109]. Plant sugars or metabolic residues occur as a result of the normal growth process and are composed primarily of the monosaccharides, glucose and fructose, and to a lesser degree the disaccharide, sucrose [110]. They may range from about 0.1–1.0% of the dry fiber weight and vary in concentration depending upon cotton fiber maturity and environmental factors (i.e., area of growth, weathering, and microbial activity). The levels of these sugars on cotton are determined by one of the several simple sugar test methods [111].

Insect sugars, commonly known as honeydew, most often come from aphids and white-flies [110]. They are found intermittently on cotton and are generally a problem with cottons grown in arid regions because rainfall in less arid regions serves to wash these impurities off the lint. Insect sugars usually are randomly deposited as spots or specks on the cotton, causing stickiness [112]. Stickiness from high levels of either plant or insect sugars on cotton lint may cause serious processing problems in textile mills, causing fibers to stick to draft rolls or other processing equipment. The sugar profiles for aphid and whitefly honeydews are quite distinct from one another. Aphid honeydew consists primarily of glucose, fructose, melezitose [113,114], maltose, and a dozen or more longer chain oligosccharides, most of which have yet to be characterized. Whitefly honeydew, by comparison, exhibits a less complex profile than the primary sugars, which are glucose, fructose, and trehalulose [115–118], a disaccharide with a high hygroscopicity. The two honeydew types are generally distinguished from one another by the presence or absence of trehalulose, as aphid honeydew contains little if any of this sugar. Arabitol and mannitol (monosugar alcohols), which are products of fungal activity, can sometimes be detected and are indicators of microbial damage to cotton [119]. The sugars are readily removed by water. They are removed by the normal scouring and bleaching processes that are used for the preparation of the fiber for dyeing and finishing.

Organic acids (0.5–1.0% of fiber dry weight) in the raw fiber, exclusive of pectic acid, are primarily 1-malic acid (up to 0.5%) and citric acid (up to 0.07%), are present in the lumen as metabolic residues, and are removed during the normal scouring and bleaching due to their high water solubility. Analyses indicate that other acids are also present, totaling some 0.3% but these have not been identified. Organic acids are removed during normal scouring.

Inorganic cations are also present as metabolic residues, again primarily in the lumen, as salts of organic acids or inorganic anions. The inorganic salts (phosphates, carbonates, and oxides) and salts of organic acids present in the raw fiber are reported as percent ash (about 1.2% of fiber dry weight) and expressed as the oxides of the elements present (excluding chlorine, which is expressed as such). The amounts of these cations present on the cotton fiber vary considerably [120] because of maturity differences, environmental factors (e.g. rainfall), and agricultural practices, as well as the field and the handling procedures that affect deposition of material (plant parts and soil) on the fiber. As is the case with all growing plants, mineral salts are necessary for the development of the cotton plant. During the production of cotton, the plant absorbs potassium and other metals as normal nutrients. Metals are incorporated from the soil into plants as natural constituents. In addition to metals absorbed by plant tissue, soil and plant parts may be deposited directly onto the lint, especially during harvesting. Ca, P, S, K, and Fe are plant part elements and Mg, Al, Si, Fe, Cr, Se, Hg, Ni, Cu, K, and Ca are soil elements [121].

The most abundant cation is potassium (2000–6500 ppm), accounting for approximately 70% by weight of total cationic content, followed by magnesium (400–1200 ppm), which accounts for approximately 14% [120]. Calcium (400–1200 ppm) is found both in the lumen and in the pectin fraction, where it serves as a cross-linking agent, and accounts for approximately 14% of the total cationic content. Sodium (100–300 ppm), iron (30–90 ppm), zinc (1–10 ppm), manganese (1–10 ppm), and copper (1–10 ppm) are also present in relatively small quantities. Lead and cadmium were not detected (Table 3.2). In untreated cotton, arsenic levels are usually less than 1 ppm [122,123]. Silicon, phosphorus, chlorine, sulfur, and boron are detected sometimes in trace amounts.

A secondary source of inorganic content on cotton fiber is the deposition of wind-borne particles onto the outer surface of the cotton fiber. Although such particles may be present only in trace amounts, their presence in cotton is of importance to processors because they can contribute to problems in yarn manufacturing, bleaching, and dyeing. Silicon as silica and other metals as oxides can cause frictional problems in rotor spinning and needle wear in knitting [120]. Iron and copper metal particles, introduced to the fiber through deposition from machinery parts, can cause problems in the peroxide bleaching process as well as contribute to a permanent coloration that may affect dyeing. Peroxide bleaching also can be affected by magnesium salts [124]. Insoluble calcium and magnesium salts can interfere with dyeing [125] and copper and iron can contribute to yellowness of the finished denim goods [126]. Iron can contribute to the permanent brown or pink color of the fiber, which affects dyeing [124]. The metals of potential concern in wastewater effluents from textile dyeing and finishing are copper and zinc. The levels of these metals in cotton fiber are low enough so that they do not contribute significantly to effluent problems [47]. The metals are removed for the most part by proper scouring and bleaching processes that are used to prepare the fiber and fabric for dyeing and finishing.

TABLE 3.2
Metal Content of Cotton

Metal	ppm
Potassium	2000–6500
Magnesium	400–1200
Calcium	400–1200
Sodium	100–300
Iron	30–90
Manganese	1–10
Copper	1–10
Zinc	1–10
Lead	n.d.[a]
Cadmium	n.d.
Arsenic	trace (<1)[b]

[a]n.d. = not detected.
[b]From Refs. [122,123].
Source: From Brushwood, D.E. and Perkins, H.H., Jr., *Text. Chem. Color.*, 26, 32, 1994.

Also of potential concern is the presence of arsenic-containing compounds, which are introduced primarily through agricultural practices such as harvest aid products (arsenic acid) and postemergent herbicides (e.g., cacadylic acid). Arsenic acid is no longer registered for use in the United States [127], and organic arsenic containing postemergent herbicides are used in less than 4% of the U.S. cotton production and are being phased out. While these compounds are generally removed through the scouring process, their presence may be of some concern for both health and marketing reasons. For solid waste, such as textile mill fiber waste (e.g., undercard and pneumafil waste), the U.S. Environmental Protection Agency (EPA) has established limits for the metals leachable from the waste. If these levels are exceeded, the waste has to be treated as hazardous [47]. Cotton does not normally contain any metals in sufficient quantity to be of concern and therefore, if the cotton fiber is not recycled, it can be disposed of in normal municipal landfills or lined landfills [47]. Some textile mill carding and other yarn manufacturing wastes are presently used as animal feed, which indicates that yarn manufacturing fiber wastes have no or very low toxicity and are generally regarded as safe.

4 Solvents for Cotton

Cellulose is soluble only in unusual and complex solvent systems. The subject has been reviewed [128–131]. Solvents for cellulose are central to the rayon and cellulose film industries, but are also necessary for solubilizing cotton for the determination of molecular weight and degree of polymerization (DP) by chromatographic methods. These solvents fall into several categories. The solvents discussed do not include processes where cellulose is converted to a derivative that is subsequently dissolved in another medium. For example, cellulose acetate is soluble in acetone, but this is not a solution of cellulose. However, the viscose process that forms a cellulose xanthate derivative, from which cellulose is readily regenerated, is generally considered to use a cellulose solution because solvation and derivatization occur simultaneously. The viscose process is the most important method for making cellulose solutions for industrial use [132]. Alkali cellulose (pulp swollen in NaOH) is pressed and aged to reduce molecular weight. Xanthation (a reaction with CS_2) takes place in a vessel that contains an inert atmosphere (CS_2–air mixtures are explosive). The orange xanthate is subsequently dissolved in aqueous alkali to make the spinning dope. The dope is pumped through spinnerets in which there are from 14 to 40,000 holes. The spun dope is converted back to cellulose by the sulfuric acid in the coagulating bath. Another system with simultaneous derivitization and dissolution uses dimethyl sulfoxide and formaldehyde [133].

Several solvents, such as cupriethylenediamine (CUEN) hydroxide, depend on the formation of metal–ion complexes with cellulose. While not as widespread in use as the viscose process, CUEN and its relatives with different metals and ammonium hydroxide find substantial industrial use [131]. The cadmium complex, cadoxen, is now the solvent of choice in laboratory work [134].

Aqueous salt solutions such as saturated zinc chloride or calcium thiocyanate can dissolve limited amounts of cellulose [131]. Two nonaqueous salt solutions with a lengthy history are ammonium thiocyanate/ammonia and dimethylacetamide/lithium chloride (DMAc/LiCl). Solutions up to about 15% can be prepared with these solvents. DMAc–LiCl has been used for molecular weight determinations of cotton [135] (see Section 1.5.2).

Trifluoroacetic acid–methylene chloride and N-methyl morpholine N-oxide monohydrate (NMMO) [136–138] are two other solvent systems that have been studied [139]. The new generic class of regenerated cellulose fibers, lyocell (e.g., Tencel [Courtaulds Fibres Limited, London, England]), is spun from aqueous solutions of NMMO [140]. Lyocell is an alternative to the generic name "rayon" for a subcategory of rayon fibers where the fiber is composed of cellulose precipitated from an organic solution in which no substitution of the hydroxyl group takes place and no chemical intermediates are formed. Lyocell may have a different crystalline structure (a mixture of cellulose II and cellulose III [141]) than other rayons and cotton cellulose. No information has been published on cotton molecular weight determinations using NMMO as the solvent.

A lengthy multiday procedure can be required to produce complete dissolution of the high-molecular-weight cellulose of cotton in the DMAc/LiCl solvent system. Other problems have been found with the DMAc/LiCl solvent system, including incomplete dissolution of some celluloses and degradation of cellulose caused by heating in the solvent system [142–145]. Despite 20 years of use of the DMAc/LiCl solvent system, a recent review [146] further documented the problems with incomplete dissolution and possible aggregation of some celluloses in DMAc/LiCl, and highlighted the need for continuing basic research on this solvent.

A new cellulose solvent system, 1,3-dimethyl-2-imidazolidinone/lithium chloride (DMI/LiCl), was recently reported [147] with the advantages of rapid complete dissolution of cellulose and reduced health risks. This solvent system was adapted for gel permeation chromatography (GPC) analysis of cotton cellulose [148] and was found to more completely, rapidly, and easily dissolve the high-molecular-weight cellulose of cotton than the DMAc/LiCl solvent system, yielding weight average molecular weights of 542400, 232900, 230700, and 114800 for cotton, ramie, linen, and Tencel, respectively. A detailed examination of the suitability of DMI/LiCl for SEC–MALLS (size-exclusion chromatography coupled to multiangle laser light scattering) analysis of cellulose found that it is a true solvent for cellulose. Cellulose molecules dissolved in this solvent are separated by their molecular mass or root-mean-square radius by SEC, no aggregates are produced, and the cellulose solutions were stable over several months at room temperature [149]. It remains to be seen if this solvent will prove to be more effective for measurements of molecular weight distributions of cotton cellulose than the DMAc/LiCl solvent.

Aqueous 9% sodium hydroxide at −5 to +5°C can dissolve steam-exploded chemical wood pulps [150]. Isogai and Atalla [151] adapted this procedure for the dissolution of microcrystalline cellulose, and compared the solubility of several other celluloses. Microcrystalline cellulose is suspended in water, sodium hydroxide is added, and the mixture is shaken at room temperature to dissolve the sodium hydroxide and suspend the cellulose in the solution. The suspension is cooled to −20°C and held at that temperature until a solid frozen mass is formed. On thawing, a gel-like mass results and with the addition of water and gentle shaking, a clear solution of 2% cellulose in 5% aqueous NaOH forms. Low-DP cellulose and Avicels (either original, regenerated, ethylenediamine (EDA)-treated, or mercerized) are 100% soluble; linters and mercerized linters are about 32% soluble. High DP fibrous plant celluloses are soluble only up to 37% even after swelling treatments.

Organic salts that are liquids at or near room temperature are referred to as ionic liquids. Ionic liquids that contain strong hydrogen-bond acceptors, such as Cl^-, Br^-, or SCN^- can be used to dissolve cellulose [152]. The authors compared the free chloride ion concentration in a typical 10% DMAc/LiCl solution, ca. 6.7%, with the chloride concentration in an ionic liquid consisting of 1-butyl-3-methylimidazolium cation and a chloride anion; in the ionic liquid, the chloride concentration is almost three times as high. They speculated that the high chloride concentration and activity in the ionic liquid is highly effective in breaking the hydrogen-bonding network of cellulose, and leads to faster dissolution and higher concentrations of cellulose than traditional solvent systems. Ionic liquids are heated with high efficiency by microwaves, and such heating speeds up the dissolution. Regenerated celluloses prepared from the above ionic liquid did not have significant changes in DP or polydispersity. Studies of the use of ionic liquids for extrusion of cellulose fibers are in progress [153].

5 Structural Properties of Cotton

CONTENTS

5.1 STRUCTURE OVERVIEW

Although there is extensive practical knowledge of cotton processing, a more thorough understanding of cotton fiber structure will improve exploitation of today's fiber. Still, the ultimate reason to seek such understanding is to be able, through genetics, biochemistry, or chemistry, to tailor fiber to have new or improved properties. The initial structure of a cotton fiber is determined by biosynthesis, a series of processes that are subject to substantial influence during fiber growth. After the boll opens, there are many factors that affect the structure, from the weather before the fiber is harvested to the industrial processes such as mercerization. Cotton fibers are composed mostly (e.g., 95%) of the long-chain carbohydrate molecule, the cellulose (the sugar of cell walls). In this overview and Section 5.2 through Section 5.5, we are concerned with the physical structure of cellulose and the fibers, as revealed by various methods.

Other commercial sources of cellulosic fibers include hemp, jute, flax (linen), and ramie. Wood fibers are used in papermaking and as a feedstock for rayon. Of these sources, cotton provides the purest cellulose. From an experimentalist's point of view, algae and even animals (the tunicates) are also interesting sources of cellulose. Bacteria such as *Acetobacter xylinium* make extracellular cellulose, but in higher plants and algae, cellulose occurs in the walls of individual cells.

Some workers refer to cotton lint (the normal fibers) as cellulose to distinguish it from seed cotton (fiber still on the seed) or the entire plant. Herein, the word cellulose has only the strict chemical meaning: linear β-(1→4)-D-glucan. In the cell wall, cellulose occurs in small, crystalline microfibrils that are arranged in multilayer structures (see Figure 5.1). An especially important layer is the primary wall (see Figure 5.2) although it is a small fraction of the mature, fully developed fiber.

Detailed structures of many crystalline materials can be determined by diffraction methods. However, because of the complex hierarchy of the cotton fiber and its very small crystallites, diffraction experiments on cotton fibers cannot provide fine details of molecular structure. Instead, the best data on cellulose structure comes from other sources. One of the major points of interest is the finding that cellulose has many different crystalline forms, or polymorphs, depending on the sources and subsequent treatments. Historically, there are four polymorphs or allomorphs, I to IV, and subclasses have been identified for all but cellulose II.

Some cotton cellulose is noncrystalline or amorphous in the sense of "lacking definite crystalline form." One reason is that cotton cellulose has a broad molecular weight

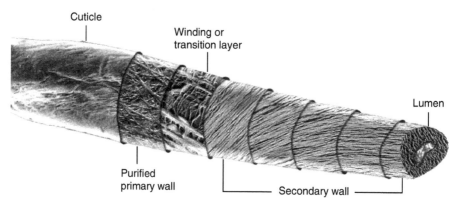

FIGURE 5.1 A computer-generated montage of a fiber segment constructed from individual transmission electron micrographs (TEMs) and scanning electron micrographs (SEMs) of layers of the fiber to show an overview of the layered structures of the fiber. Fiber surface, primary wall, and secondary layers have been shown at different magnifications to better visualize fibrillar structures and the various fiber layers from surface to lumen. The surface marked cuticle is an SEM view of a scoured and bleached fiber surface and was used as the skeleton of the montage. All other segments are taken from transmission micrographs and are shown at higher magnifications. No fibrils are visible in the SEM of the cuticle because of its relatively low magnification as well as the presence of noncellulosic materials. The fiber surface at the cut end of the fiber shows fibrils that have been separated by swelling. Although the montage uses actual pictures of cotton fiber layer structures, it is intended to represent a possible fiber morphology rather than to indicate an exact cotton fiber structure. (Credit to Wilton Goynes.)

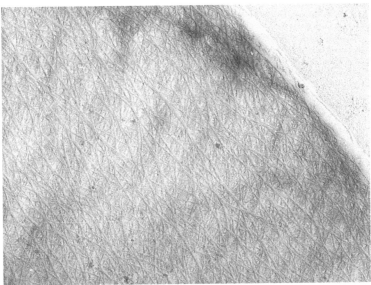

FIGURE 5.2 Transmission electron micrograph of the loose tangle of cellulose microfibrils in a cotton fiber cell wall. (Credit to Wilton Goynes.)

distribution, making high-crystalline perfection impossible. The small crystallites constitute deviations from ideal crystals that are infinite arrays. The remaining amorphous character of most polymers is often thought to arise from the "fringed micelle" model of the solid structure. In that model, the polymer molecules, much longer than the individual crystallites, pass through several crystallites (micelles). Amorphous regions exist where the molecules are not part of any crystallite.

Evidence for periodic interruption of the crystalline regions comes from the leveling-off degree of polymerization (LODP) studies. In such work, the cellulose is hydrolyzed in 1 M HCl at 55°C. The molecular weight drops rapidly for about a day, after which the drop is much slower. The interpretation is that periodic amorphous regions are readily attacked by the acid but the crystalline regions are much more resistant. After the acid treatment, the remaining material, called hydrocellulose, is more crystalline than the starting material. The LODP values for cotton are about 175 glucose units [154], whereas ramie cellulose values are about 300 [155]. In the latter work, small angle neutron scattering of lightly deuterated samples confirmed the spacing.

Other amorphous contributions arise from imperfections in the crystals. These defects include molecular ends inside the microfibrils, causing discontinuities, disorder on the crystallite surfaces (which may comprise as many as half of the molecules in these small crystals), and mechanical bending and twisting of the crystallites [156]. These bases for the amorphous character are embodied in Figure 5.3. This representation is attractive because of the high density of fibrous cotton (e.g., 1.55 g/cm^3) when purely crystalline cellulose is 1.62 g/cm^3. The high density suggests that the chains are more orderly and compactly arranged than indicated by most depictions of fringed micelles.

Given a DP of about 20,000 glucose residues, a stretched out molecule is perhaps 100,000 Å (10^{-5} m) long. Thus, compared to the fiber's length of about 30 mm (3×10^{-2} m), a cotton fiber is some 3000 times longer than a molecule. Coincidentally, the molecular length of 10 μm is roughly the same as the diameter of the cotton fiber. In fibers such as linen, the molecules are aligned parallel to the fiber axis to within a few degrees. In cotton fibers,

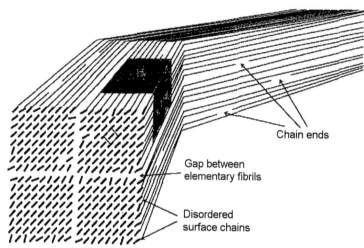

Chain ends

Gap between
elementary fibrils

Disordered
surface chains

FIGURE 5.3 Cellulose microfibril composed of four elementary fibrils, viewed along the chain axes. The unit cell is shown, and the chains on the elementary fibril surfaces are disordered. Chain ends within the elementary fibrils account for more disorder. The surfaces of the crystallites are related to the crystallographic planes shown in gray.

however, alignments are distributed over a much larger range. Early in the fiber development, the fibrils are parts of, or near, the primary wall and some have angles nearly perpendicular to the fiber axis (See Figure 5.2). As the secondary cell wall develops on the interior of the primary wall, the microfibrils are thought to be progressively aligned more closely with the fiber axis (Figure 5.1). Within a given layer of secondary wall, the direction of the molecules may be roughly constant until it suddenly changes, causing a reversal to appear. Cellulose is birefringent and these reversals are visible under a polarizing optical microscope as well as in the scanning electron microscope (SEM) (see Section 5.5).

The mature cotton fiber has a noncellulosic covering called the cuticle that contains waxes, pectins, and proteins left over from biosynthesis (see Chapter 3). This cuticle is intermingled with the primary wall. Figure 5.4 shows the location of immunolabeled pectin within the primary wall. The structure of the primary wall, which changes substantially during fiber development, is not well understood. It is responsible for maintaining the integrity of the fiber and may account for much of the strength of the cotton fiber. Most of the cuticle is dissolved and removed by industrial scouring of fabric, but it has important functions during spinning of the fibers into yarn and during weaving the yarn into fabric. One reason for scouring is that the waxes block access to the interior of the cotton fiber for molecules such as dyes.

Thus, a complete description of fiber structure requires knowledge of the structure of the cellulose molecule, the structure and perfection of its crystalline arrays, the packing of these arrays (elementary fibrils) into microfibrils, and then the arrangement of these microfibrils in the primary and secondary cell walls. The noncrystalline material is also important. The large-scale structures must also be understood. Fibers have various pores or voids that are important in textile finishing reactions. After the cotton boll opens, the cellular fluid dries, leaving a cavity, the lumen, which contains biological material. That material constitutes a small percentage of the total dry weight. With drying of the fiber, the lumen collapses, leaving fiber cross-sections with irregular, kidney-bean shapes, as shown in Figure 5.5. Chapter 7 discusses how the degree of development (thickness and shape) of the secondary wall (the maturity) affects many performance characteristics.

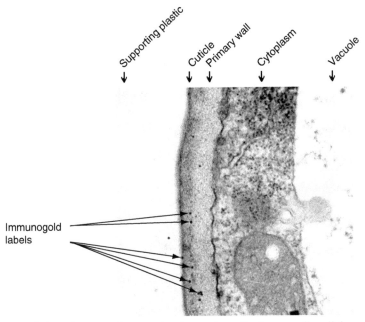

FIGURE 5.4 Transmission electron micrograph of a section of primary cell wall from very immature cotton fiber. The black dots are immuno-labels of pectin molecules. (Credit to Kevin Vaughn, USDA-ARS, Stoneville, MS.)

Treatments of cotton fiber often alter the structure at more than one level. Therefore, it is not always obvious that the effect of the treatment should correlate with a change at a particular level of structure. For example, cotton is often treated with NaOH in a process called mercerization. This treatment can alter the crystal structure, but it also reduces the density of the whole fiber and increases its luster. For example, the change in crystal structure is probably not directly responsible for the change in fiber luster.

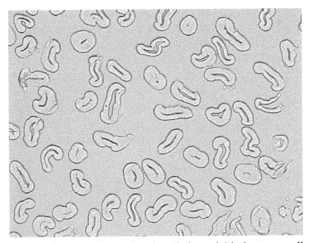

FIGURE 5.5 Cross-sections of cotton fibers showing their variable but generally kidney-bean shapes and lumens. These fibers are from a lot of cotton that, on average, was fully mature. (Credit to Devron Thibodeaux.)

5.2 CELLULOSE MOLECULE—CONSTITUTION AND MOLECULAR WEIGHT DISTRIBUTIONS

5.2.1 CONSTITUTION OF CHAIN

Evidence from degradation by hydrolysis, oxidation, and other chemical reactions shows that cellulose is a 1→4-linked linear polymer of β-D-glucopyranose (Figure 5.6). These monomers are linked together by elimination of one molecule of water between the hydroxyl groups attached to the number 1 carbon atom of one glucose molecule and the number 4 carbon atom of another about to be joined with loss of the circled (Figure 5.7). Repetitions of these condensations during biosynthesis (with a more complicated pathway, to be sure) lead to unbranched polymer chains of great lengths (Figure 5.8). In undegraded cotton fiber, the molecular chain length (degree of polymerization [DP]) may be higher than 20,000 monomeric D-glucopyranosyl units. This corresponds to a molecular weight of 3,240,000 Da. Because the condensation reaction to make the cellulose polymer eliminates a molecule of water per glucose, the monomers are often referred to as anhydroglucose units or as glucose residues. Each cellulose chain has a reducing end (O1—H) and a nonreducing end (O4—H). Reducing ends are especially reactive, but they are present in such small amounts in cellulose that they are often ignored.

It is often stated that the repeating unit of cellulose is a cellobiose residue, a hydrolysis product of cellulose that is composed of two glucose residues. This acknowledges that cellulose and cellobiose both have β-(1→4)-glucosidic linkages, and that the cellulose chain has twofold screw-axis (denoted 2_1) symmetry in a number of different crystalline forms. This symmetry gives each successive residue a 180° rotation about the fiber axis and a translation of half the crystallographic repeat of ~10.35 Å (1.035 nm). However, we suggest that it is more useful to designate a glucose residue as the repeated unit. In common usage, DP values and symmetry descriptors apply to glucose residues, not cellobiose. Cellulose can take other shapes, and it can be argued that it is not appropriate to define a flexible molecule by only one of its shapes. Interestingly, cellobiose itself cannot have intramolecular 2_1 symmetry. Finally, it is just as necessary to specify that cellobiose units are joined to each other with β-(1→4)-linkages as it is for glucose units.

5.2.2 MOLECULAR WEIGHT DISTRIBUTIONS

It has long been thought that cotton fiber strength is influenced by the structural organization of the cellulose chains [157]. Molecular weight of a polymer is one of the most important influences on its physical properties, and the determination of molecular weight distribution is critical for predicting performance of a polymer. Vastly differing distributions can produce the same averages but physical properties will reflect the disparities. For polymers, higher molecular weight and narrower molecular weight distribution are positively correlated with increased strength. Unfortunately, polymer characterization techniques generally depend upon dissolving of the polymers. Attempts to identify the true molecular weight of native cellulose have been limited mostly because cellulose is difficult to dissolve (see also Chapter 4).

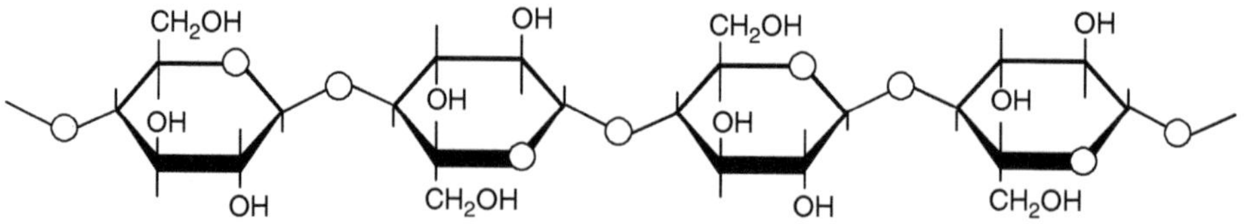

FIGURE 5.6 Chemist drawing of four-residue segment of cellulose chain.

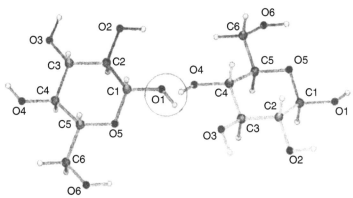

FIGURE 5.7 Two glucose residues in the $^4C_1$ (chair) conformation about to be joined, losing the circled hydroxyl group and hydrogen atom. Drawings were made with Chem-X, formerly developed, and distributed by Chemical Design, Ltd.

In addition, cellulose solutions are often unstable and sometimes undergo rapid oxidative degradation. Alternatively, cellulose can be converted to a derivative that is soluble in an organic solvent, a process that can also degrade the molecule.

In 1948, Hessler et al. [157] measured the DP of cotton based on the viscosity of nitrated cellulose. They found a DP of 5940 for the primary wall, while a DP of 10,650 was determined for the secondary wall. Molecular weight differences were reported for fiber taken from three different positions within the boll as well as at three different positions on the seed

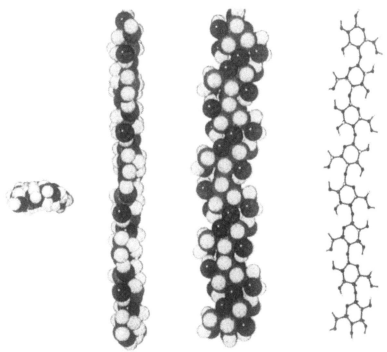

FIGURE 5.8 Projections of short cellulose chains. Leftmost: along chain axis. Left-center: along ribbon edge. Right-center: Maximal width view. Rightmost: Ball and stick model in the same orientation as the space-filling model to its left.

for three different varieties of cotton. Losses in molecular weight were observed for cotton fiber exposed to weathering in the field. Despite the strong arguments by this group for relationship between the molecular structure and the physical property (strength) of the cotton fiber, only occasional measurements of molecular weight were subsequently reported, probably because of the experimental difficulties. Marx-Figini [158] fractionated various cellulose derivatives and determined molecular weights and distributions. Her work showed for cotton fiber:

1. Secondary wall cellulose has a much larger molecular weight than primary wall cellulose.
2. The nearly constant DP of secondary wall cellulose and variability of molecular weight in the primary wall during fiber development.

More recently, nondegrading solvents for cellulose have been employed for character-ization. DMAc/LiCl [159–161] is particularly useful because there is no degradation of the polymer by the solvent, in direct contrast to other cellulose solvents that rapidly degrade the macromolecular backbone [162]. DMAc/LiCl also dissolves proteins, other polysacchar-ides, and their derivatives. The GPC can separate macromolecules of high molecular weight with the advantage that the molecular weight distribution of a polymer can be obtained in a relatively short time using automated, computer-based data acquisition and calculations. Multiple detectors bypass the need for cellulose standards (which are not available).

Cotton fibers dissolved directly in the solvent DMAc/LiCl have been analyzed by GPC [94]. This procedure completely solubilizes wall polymers without extraction or derivatization for GPC separation processes that could degrade high-molecular-weight components. This approach permits the characterization of the entire array of cell wall polymers, not previously possible, and represents a step toward solving the major problem of precise analysis. The molecular weight distribution for mature, field-grown cotton fiber from a genetic standard variety (*Gossypium hirsutum*, Texas Marker-1) is shown in Figure 5.9. Again, the identified locations of the primary and secondary walls showed:

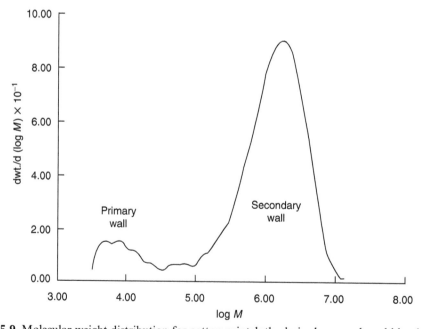

FIGURE 5.9 Molecular weight distribution for cotton printcloth, desized, scoured, and bleached.

1. Lower molecular weight for the primary compared with the secondary walls
2. The larger weight fraction of material found in the secondary wall, confirming limited previous reports [157,158]

During development, the composition of the cell wall of the cotton fiber changes continuously, ending with the cessation of the fiber's metabolic activity [163]. Cell wall polymers from cotton fibers under developmental stages were characterized via GPC analysis of DMAc/LiCl solutions [94]. Primary and secondary wall compositions of cotton fiber polymers were monitored from 8–60 DPA. As expected, cell wall polymers from fibers at primary cell wall stages had lower molecular weights than the cellulose from fibers at the secondary wall stages. However, the high-molecular-weight cellulose characteristic of mature cotton was detected as early as 8 DPA. High-molecular-weight material decreased during the period of 10–18 DPA with a concomitant increase in lower-molecular-weight wall components, possibly indicating hydrolysis during the later stages of elongation. During the maturation stages (past about 20 DPA), the high-molecular-weight components of the secondary wall increased dramatically to give the fiber profile shown in Figure 5.9. These observations are consistent with a picture of fiber development that starts with the construction of the full-sized primary wall and proceeds through the addition of secondary wall material on the inside of the primary wall.

Molecular weight distributions were determined for three commercial varieties of American Upland cotton that were similar in all physical properties except strength as measured by the high volume instrument (HVI) [164]. (It is an unsolved mystery why traditional Stelometer strength measurements gave almost identical values for all three varieties.) Fiber samples from the three varieties had different molecular weight distributions, different locations of peaks for the secondary wall fraction, and different weight average molecular weights. Ranking by HVI strength corresponded with ranking by peak molecular weight and weight average molecular weight. Fiber classification standards (for HVI) representing a range of lengths and strengths were sampled and assessed by GPC analysis [165]. The shortest fiber (0.903 in.) with a Stelometer strength of 21.4 g/tex had DP_w (Weight averaged degree of polymerization) = 15,000. The longer fiber (1.236 in) with 40% higher strength (31.0 g/tex) had DP_w = 23,700. The relation of the average DP_w to strength is shown in Figure 5.10.

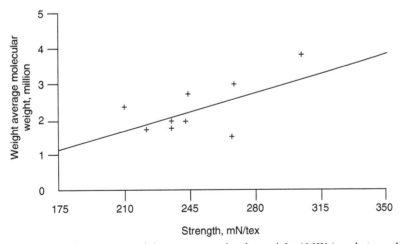

FIGURE 5.10 Relationship between weight average molecular weight (MW_w) and strength for cotton fibers with a range of lengths and strengths. Line shown is least squares fit with R of 0.62.

The general correlation is evident between molecular weight and strength whereby the weight average molecular weight accounts for ~38% of the variability in strength. Thus, molecular compositional profiles indicate correlation of higher average molecular weight with greater strength of the cotton fiber. However, exceptions were evident. Two samples with equivalent lengths and strengths had differences in weight-average molecular weights. The sample with the higher molecular weight had, however, a broader-molecular-weight distribution than the lower-molecular-weight sample. The increased molecular weight in one sample apparently was offset by the narrowness of the molecular weight distribution in the other, giving equivalent strengths.

Molecular weight distributions of cotton fibers were evaluated with variables of the variety and the growth environment [166]. GPC analysis of cotton fiber samples demonstrated consistency of the secondary wall cellulose peak for the variety, alterations in the composition of the cellulose chains as a response to dryland conditions, and differences in molecular weight distributions according to fruiting zones for different irrigation methods.

After cross-linking cotton fabrics with dimethylolethyleneurea to impart durable-press (nonwrinkling) properties, the cellulose was evaluated for strength losses and structural changes [167]. Cellulose nitrate derived from the cotton fabric before and after cross-linking treatment was dissolved in tetrahydrofuran for analysis by GPC. In this case, nitration is advantageous because the cross-links are broken during the derivatization thereby facilitating the characterization. Molecular degradation and cross-link embrittlement were measured as a function of treatment conditions. In fabrics with shorter cure times, the predominant strength loss (which dropped rapidly) came from cross-link embrittlement. Molecular degradation became the major source of strength loss at longer cure times. Acid-catalyzed treatment of the fabric with the substituted ureas substantially reduced the DP of the cotton cellulose.

Molecular weight distributions of cotton fabrics were compared before and after treatment with a total cellulase [168]. A lengthy multiday procedure was required to completely dissolve the cotton in the DMAc/LiCl solvent system. Despite considerable weight loss and breaking load reduction produced by the cellulase treatment, GPC analysis using DMAc/LiCl did not show reduction in molecular weights of the cellulose. This result supports a hypothesis that exoglucanases rapidly cleave cellobiose units from cellulose chains once the chains have been clipped by endoglucanases, resulting in a total removal of cellulose chains and a reduction in microfibrillar size, but with the remaining chains not yet degraded by the total cellulase. GPC analysis of DMAc/LiCl solutions of enzymatically treated lyocell made possible a clarification of depilling mechanisms [169]. The relatively low-molecular-weight cellulose of lyocell was rapidly and easily dissolved, in contrast to the multiday dissolution of the high-molecular-weight cotton cellulose noted above.

Incomplete dissolution of a sample results in an overestimation of the molecular weight. GPC analysis of cotton fabric using the DMI/LiCl solvent system appears to yield a faster, more complete dissolution of high-molecular-weight cellulose than does DMAc/LiCl. It gave a weight average molecular weight of 542,400 $\pm$ 23,200 for cotton printcloth [148].

5.3   THREE-DIMENSIONAL STRUCTURES OF CELLULOSE MOLECULES, CRYSTALLITES, AND FIBERS

Near the end of Section 5.1, there was a lengthy list of attributes needed to specify the three-dimensional structure of cellulose. The present section is concerned with descriptions of the smaller-scale structures. The information on shape comes from both theoretical and experimental methods, with the classic advantages and disadvantages of each. Theoretical methods can be used to fill the gaps between experiments and help interpret ambiguous results, but in

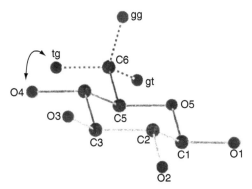

FIGURE 5.11 β-D-Glucopyranose with all three staggered orientations of O6 shown (*gg*, *gt*, and *tg*).

the end, they are still "educated guesses." On the other hand, laboratory experiments on cellulose give real data, but they are done on single, specific situations. These results may not apply in general and are often ambiguous. It is much more powerful to combine both approaches.

The shapes of cellulose chains are consequences of the slightly variable shape of the glucose ring and the more variable geometry of the β-(1→4) linkage, along with some environmental influence. Almost all D-glucose rings have the chair shape ($^4C_1$) that places the bonds to the hydroxyl and hydroxymethyl groups, including O1 and O4, in equatorial (nearly parallel to the main plane of the pyranosyl ring) orientations (Figure 5.11). We are accustomed to the chair form of the ring, but the preferred linkage geometry is a more involved question, discussed below.

5.3.1 DESCRIBING MOLECULAR SHAPE

When all monomeric units and their linkages in a polymer have identical geometries, the molecule will meet the mathematical criteria for a helix. A molecule need not have a central cavity to be helical; it suffices to have intramolecular screw-axis symmetry. Helices are described in terms of the number n of monomeric units (glucose residues) per turn of the helix, and whether the screw thread is left- or right-handed. If left-handed, n is given a minus sign. In case of $n = 2$, however, the helix can be considered to be of either chirality. Also, the advance along the helix axis per residue h and the pitch P the distance between turns of the helix, signify how extended or collapsed the helix is. The cellulose chain is mostly extended, even when it is not in a crystal (see Figure 5.8). For common crystalline cellulose, these parameters are $n = 2$ and $h \approx 5.18$ Å. The degree of extension can be understood by comparing h with the length of the glucose unit in small-molecule crystals, 5.42 to 5.57 Å [170]. Twofold screw helices have symmetry such that for every atom at x, y, and z, there is an identical atom at $-x$, $-y$, $(z + 0.5c, c = P)$. The x and y values are distances from the helix axis, and z is the distance along the crystal unit cell of dimension c and the fiber axis. This definition in Cartesian coordinates is an alternative but equivalent definition to the one in Section 5.2.1 that was in cylindrical polar coordinates.

If not in a crystal, cellulose chains would deviate from the ribbon shape in a random way, dictated by the lowest free energy. Deviations from the shape that has the lowest possible enthalpy do not increase the enthalpy very much, but variation increases the entropy. Therefore, deviations are favored. Still, even in solution or noncrystalline regions, the cellulose chain is expected to be rather extended, with the occasional kink [171]. In any case, helix nomenclature can still apply, but only in approximation for local segments.

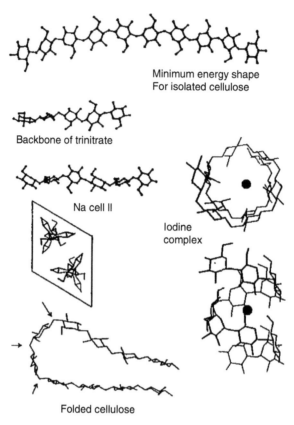

Minimum energy shape
For isolated cellulose

Backbone of trinitrate

Na cell II

Iodine
complex

Folded cellulose

FIGURE 5.12 Various shapes of cellulose backbones proposed [18]. The arrows near the hairpin structure point to glycosidic linkages having approximate conformations of $\phi = 0°$ and $\psi = 180°$.

Crystalline cellulose derivatives take other shapes, such as the trinitrocellulose helix with five residues repeating in two turns with $n = |2.5|$ [172]. The nitromethane complex of cellulose triacetate has eight residues in three turns ($n = -2.67$) [173]. Along with numerous cellulose derivatives that have $n = |3|$, the soda cellulose II complex is a threefold helix that is probably left-handed [174,175]. Values of h are still greater than 5 Å, indicating substantial extension. Greater deviation from the twofold structure is also possible. According to energy calculations, loop structures (Figure 5.12) are also possible, although these are usually somewhat higher in energy than the simple extended models. These other shapes raise the question as to whether the 2_1 shape observed in pure cellulose crystals is the favored form or whether it is distorted so it can pack most efficiently. This is a good opportunity for input from modeling.

5.3.2 EXPERIMENT- AND THEORY-BASED MODELING STUDIES

One way to study the shapes of cellulose chains is to construct models that accommodate the available experimental data. There are many approaches to modeling, and comprehensive studies require extensive computations. The first computer model of a carbohydrate was a part of the experimental diffraction studies of cellulose [176]. Since then, there have been substantial improvements in both computers and their representations of molecules.

5.3.2.1 Extrapolated Experimental Models

The simplest models are extrapolations [170] of the experimentally determined structures of small molecules that are related to cellulose, such as cellobiose. Many such structures have been determined and are conveniently available in large databases [177,178]. The models consist of the x, y, and z coordinates of the atoms. Given the coordinates, it is simple to calculate the bond lengths, bond angles, and torsion angles. Purely geometric manipulation is used to repeat the geometry of a single experimentally observed glucose unit, connected by an observed linkage geometry. Instead of constructing models from each combination of glucose and linkage geometry, it is useful to take a shortcut. This shortcut consists of a generalized conversion from the ϕ and ψ torsion angle values to n and h values, using an average glucose ring geometry and glycosidic angle, τ (C1'—O4—C4). Figure 5.13 shows the ϕ and ψ torsion angles and valence angle, τ, which are used to characterize the linkage geometry. The values of ϕ and ψ range from 0 to 360° and are calculated from the O5'—C1'—O4—C4 and C1'—O4—C4—C5 torsion angles. Our conversion uses a constant value of 116° for τ. A conversion chart from ϕ and ψ to n and h is shown in Figure 5.14; its axis values are chosen so that the $n = 2$ line is a centered diagonal.

For example, take the atomic coordinates of the nonreducing ring of β-cellobiose [179] and the linkage geometry of that same structure. Exact extrapolation gives a helix with $n = -2.38$ and $h = 5.13$ Å [170]. Similarly, the hydrated sodium iodide complex of α-cellobiose [180] leads to helices with $n = -2.82$ and $h = 5.04$ Å, and one crystalline form of polymorphic methyl-4-O-methyl-β-cellobioside [181] yields helices with $n = 2.04$ and $h = 5.14$ Å. Review of the small filled circles in Figure 5.14 shows that the latter two values of n cover most of the range of extrapolations of small-molecule crystal structures, with one exception. That is a heavily substituted cellobiose [182] with $\phi = -71°$ and $\psi = -292°$. It would lead to helices with $n \approx +4.5$ and $h \approx 3.1$ Å. With that exception, the extrapolated chains all have $h = 5$ Å or greater and $n = 2$ to 3, with most of the chains left-handed. Approximate twofold conformations exist even when no O3'—H···O5 hydrogen bond or internal symmetry is possible, such as for acetates of xylobiose [183] and cellotriose [184].

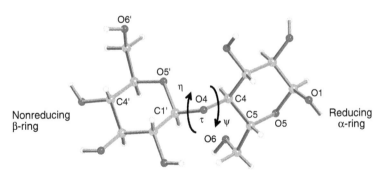

FIGURE 5.13 Cellobiose with the locations of the important shape variables ϕ, ψ, and τ. τ is a conventional bond angle, C1'—O4—C4, and ϕ and ψ are torsion angles that specify the amount of twist about the C1'—O and O4—C4 bonds. ϕ and ψ can be defined by any of the three atoms attached to C1' or C4, respectively, and perhaps the most often used definitions are $\phi = $ H1'—C1'—O4—C4 and $\psi = $ C1'—O4—C4—H4. Recently, we have been using $\phi = $ O5'—C1'—O4—C4 and $\psi = $ C1'—O4—C4—C5 because we compare our energy surfaces with data from crystal structures in which the hydrogen atoms are often poorly located, even though the carbon and oxygen atoms are accurately determined. O6 atoms are in the gg position in this α-cellobiose structure taken from Peralta–Inga et al. [177].

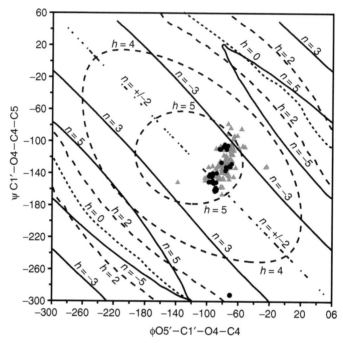

FIGURE 5.14 Contours of *iso-n* and *iso-h* values in $\phi-\psi$ space for model cellulose helices composed of the nonreducing residue of crystalline cellobiose [179] and a value of $\tau = 116°$. Also shown as dots are the experimentally determined values of ϕ and ψ from crystal structures of small molecules related to cellulose. With one exception at the bottom of the map, all of the dots correspond to extended helices with the numbers of residues per turn between 2 and 3. Geometries found in complexes of cellulose fragments and proteins are shown as triangles. They have a similar distribution but the range is expanded. Left-handed helices are indicated with negative values of *n*.

Somewhat larger variations in model cellulose shapes are found if geometries are extrapolated from the less-accurately determined crystals of proteins that are complexed with fragments of cellulose or related molecules. There are more than 100 values of the linkage geometry for such complexes in the literature, shown as triangles in Figure 5.14. As described above for the small-molecule crystals, there is a heavy concentration of observed values of ϕ and ψ that correspond to *n* and *h* values between 2 and −3. Unlike the results from the simple crystals, some of the points on Figure 5.14 could be in error. Other points are intriguing because they may indicate that the protein is causing distortions that have a beneficial purpose, such as increased catalytic activity.

5.3.2.2 Theoretical Models

The above extrapolated shapes for cellulose depend on the availability of observed structures. They also depend on the assumption that a geometry that is observed in one environment could exist under other conditions. A different modeling approach is to calculate the internal, intrinsic energy of all possible structures. The probability would be maximal for the structure with lowest energy and then would decrease according to the higher energy values for the remaining structures. This can be understood by comparing the molecule to a spring. If a spring is stretched or compressed from its lowest energy (relaxed) shape, then it will have potential energy to restore itself to the lowest energy form when the compression or stretching force is removed.

Energy calculations are based on two fundamentally different methods. One of the methods, the electronic structure theory describes a molecule by approximating its electron orbitals with probability functions and makes use of very fundamental physical constants. Such *ab initio* quantum mechanics (QM) calculations improve the accuracy, as more complete theory is used. However, the advanced levels of theory require a great deal of computer time. Alternatively, the energy can be calculated with empirical force fields, also called molecular mechanics (MM) [185]. Here, components of molecular structure such as bond lengths, bond angles, torsion angles, van der Waals forces, and electrostatic forces, are described with simple equations that can be evaluated quickly. For example, MM calculations of energy for changes of bond length can be modeled with Hooke's law for stretching a spring. There are varied approximations used in these force fields, and variations in results from different methods and implementations are the basis of continuing controversy.

Either MM or QM can be used to carry out energy minimization. For example, a molecule can be drawn and its crude coordinates can be submitted for geometry optimization. The modeling software would systematically shift the atoms until the calculated energy was minimized. However, there is no way to know that this local minimum is the lowest possible (global) minimum. The method of conformational analysis systematically puts the molecule into all possible shapes and, in recent times, minimizes the energy at each increment of change.

Besides energy minimization, molecular dynamics (MD) can be employed. In models using molecular dynamics, the individual atoms have kinetic energy commensurate with their temperatures. The atomic motion resulting from the kinetic energy is restrained by the potential energy, depicting variations in molecular shape over time. The individual structural components, such as a particular torsion angle, may be monitored and this record of the time-based variation is called a trajectory. The MD is also used for conformational searching of complex molecules, especially at higher temperatures. It permits explicit inclusion of solvent in the model, and is advantageous because solutions are inherently dynamic. Understandably, the computer time needed increases substantially when hundreds or thousands of solvent molecules are included.

Cellobiose, the shortest cellulose chain, has been extensively modeled with conformational analysis and MD. When solvent water is present, it competes with the cellobiose hydroxyl groups [186]. The Chemistry at Harvard Macromolecular Mechanics (CHARMM) program found a new region especially favored by solvated cellobiose [187]. QM has also been used to find the lowest energy form of cellobiose [188], which, if extrapolated as above for a few residues, would lead to folding. A conformational analysis using QM of a cellobiose analog that is missing all of the hydroxyl and hydroxymethyl groups is reasonably predictive of the observed shapes (see Figure 5.15), even though many of the observed shapes are from structures that have interring hydrogen bonds.

Hybrid modeling studies of cellobiose that are based on both the QM calculations (Figure 5.15) and MM4 for the hydroxyl and hydroxy methyl groups are even more predictive of the observed shapes of the cellobiose linkage [189]. That was true only as long as the electrostatic forces are substantially reduced compared to the full strength. Nonexperimental molecular conformations were apparently favored when using potential energy functions that have strong hydrogen bonding [188,190].

Longer cellulose oligomers have been modeled with MD as well. Two studies have attempted to discuss the molecular shapes in aqueous solution as well as the solvent–cellodextrin and cellodextrin–cellodextrin interactions [191,192]. The first of these studies showed that chains were heavily solvated and not fully extended or in contact with other cellulose fragments. The simulation was proposed as a model for freshly prepared cellophane. The latter study was more in agreement regarding chain shapes with the results in Figure 5.16. In addition, that work showed, unlike the simulated cellotetraose molecules in the same study,

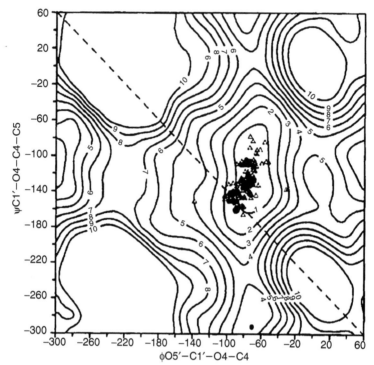

FIGURE 5.15 Quantum mechanics (B3LYP/6-311++G**) energy surface for a cellobiose analog composed of tetrahydropyran residues (cellobiose with all of the exocyclic groups replaced by hydrogen atoms). Contours are shown in 1 kcal/mol increments up to 10 kcal/mol of relative energy. Dots indicate the ϕ and ψ values of experimentally determined structures of small molecules and triangles represent comparable values of structures of molecules related to cellulose that are in complexes with crystalline proteins. The diagonal line represents conformations that would have twofold screw-axis symmetry.

that cellohexaose fragments stayed in contact with each other during the simulation. That was in agreement with the low solubility of longer chain crystals. Earlier, MD studies of cellooctaose showed that the central residues in such long chains might undergo major changes in ring shape [193]. Such deformations might be found in the amorphous regions of fibrous materials. Work to model the mechanical properties of cellulose was based on MM, with proposed differences in intramolecular hydrogen bonding for cellulose I and cellulose II leading to computed moduli similar to experimental values of about 130 and 80 GPa (GN/m^2) [194].

5.3.3 EXPERIMENTAL STUDIES ON CELLULOSE STRUCTURE

Now that the range of likely shapes has been defined by experiments on related molecules and by energy calculations, we focus on the details of specific structures that have been observed for real, crystalline cellulose molecules, primarily by x-ray, neutron, and electron diffraction studies. A number of landmark concepts have been established with electron microscopy, as well. Infrared (IR), Raman, and nuclear magnetic resonance (NMR) spectroscopy have all also been important in the quest for understanding cellulose structure. Such data, while so far not able to provide complete definitive structures themselves, constitutes additional criteria that any proposed structure must be able to explain. In addition, unlike crystallography, the resolution of spectroscopic methods is not directly affected by the dimensions of the

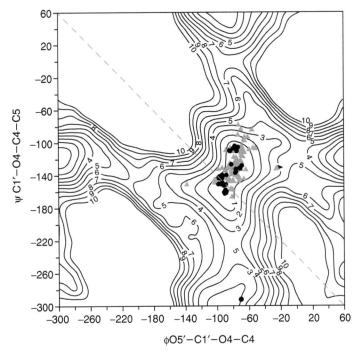

FIGURE 5.16 Energy map over $\phi-\psi$ space for cellobiose based on the quantum mechanics energies from Figure 5.15 for the backbone and empirical force field (MM4) energies for the hydroxyl and hydroxymethyl groups.

crystalline units, but only by the degree of order within the crystallites [195]. A comprehensive review of the contributions of spectroscopy to cellulose structure has been provided by Rajai Atalla [196].

5.3.3.1 Basic Information on Diffraction

Diffraction occurs when a beam of radiation with a wavelength comparable to atom–atom distances interacts with the periodic arrays of molecules in a crystal. Diffracted rays are recorded on imaging systems of many different types, originally photographic film. The positions of the diffracted rays indicate the dimensions of the unit cell, the smallest unit of the crystal structure that can be repeated by simple translation to generate the entire crystal. The intensities of the diffracted rays indicate the types and relative positions of the constituent atoms, and the reciprocal of the width of the diffracted ray indicates the size and degree of perfection of the crystal. Radiation wavelengths of around 1 Å are provided by x-ray generators, electron microscopes (electron beams), and nuclear reactors (neutron beams). Synchrotrons produce very intense x-ray beams.

Cellulose samples for diffraction studies include powders, fibers, and films, all of which are more difficult to analyze than the relatively large single crystals formed by many small molecules and even proteins. A single crystal of a sodium iodide complex of cellobiose gave a total of 9474 unique spots. Oriented fiber (Figure 5.17, top left and top right) and film sample usually have all the crystallites orientated about a single axis, giving much less information than single crystals. Powders consist of many small crystals that are randomly oriented, and therefore give a pattern of concentric rings. These rings are just extensions of the arcs shown

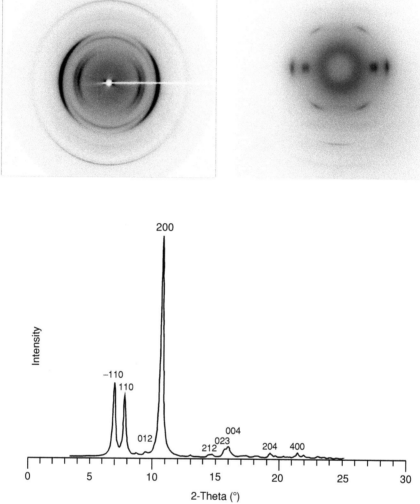

FIGURE 5.17 Cellulose diffraction patterns. Top left: synchrotron radiation x-ray diffraction pattern for cotton fiber bundle. The fiber was vertical and the white circle and line correspond to a shadow from the main beam catcher and its support. (Credit to Zakhia Ford.) Top right: electron diffraction pattern of fragments of cotton secondary wall. The much shorter arcs in the top right figure are due to the good alignment and small number of crystallites in the electron beam. (Credit to Richard J. Schmidt.) Bottom: a synthesized powder pattern for cellulose, based on the unit cell dimensions and crystalline coordinates of Nishiyama et al. [209]. (Credit to Zakhia Ford.) Also shown are the *hkl* values for the Miller indices. The 2-theta values are for molybdenum radiation instead of the more commonly used copper radiation.

in Figure 5.17 (left). Often, the collected data (Figure 5.17, bottom) is shown as a trace from the pattern center, through the rings. Imagine for example that the arcs in Figure 5.17 (top) were continuous circles, and the white shadow from the main beam catcher (the small white circle at the center of Figure 5.17, left) is at 0° 2-theta. The white line from the shadow of the beam catcher support might then represent the trace through the intensities that is shown in Figure 5.17 (bottom).

Historically, cellulose diffraction was important in discovering the characters of polymeric materials. At first, it was a great puzzle to understand how such a long molecule could fit into a unit cell that is large enough to hold only four glucose residues. The mystery was resolved when it was realized that each polymer molecule passes through many unit cells, and that the reducing and nonreducing ends can be ignored because, relatively speaking, their numbers are very small. This approximation required a shift in thinking from that needed for the non-polymeric structures that were the subjects of most studies at the time.

Cellulose crystallizes in various forms (allomorphs or polymorphs) I to IV, with subclasses. Recently, our knowledge of these different crystal structures has advanced because of studies of highly crystalline cellulose films with synchrotron x-ray and neutron diffraction. The neutron data enabled the determination of the hydrogen bonding systems. The molecular shape and general mode of packing in these different forms is similar. As a result of crystal packing, the chains take shapes that are flat ribbons in cross section, which can be packed with high density (as high as 1.62 g/cm^3). This packing results in low free energy because of the extensive, intermolecular interactions. The wider (about 9 Å), flat sides of the ribbons are hydrophobic while the thin (about 4 Å) edges of the ribbons are covered with hydroxyl groups. These chains are placed with their flat sides next to each other, allowing extensive interchain van der Waals interactions. The hydroxyl groups on the thin edges are similarly adjacent, and can form intermolecular hydrogen bonds. With such efficient, high-density arrangements, it is not surprising that cellulose is difficult to dissolve. A solvent system must disrupt both the hydrogen bonding and the van der Waals forces to break up the crystalline solid. Although chemically pure cellulose can quickly recrystallize, occasional substitution of a hydroxyl group by a methoxyl group prevents the cellulose chains from fitting well in a crystal lattice. Therefore, partially substituted cellulose chains are much more soluble than pure cellulose.

The various data peaks are referred to by their Miller indices (h, k, and l), which are tied to different periodic spacings in the unit cell (Figure 5.17, bottom). Historically the second setting for the $P2_1$ space group was preferred. It used b as the fiber axis and β as the monoclinic angle. Many reports of the unit cell dimensions and crystallite sizes are in terms of that convention. Another point of confusion is that earlier work used monoclinic angles (β or γ) less than 90°. The convention recommended by Klug and Alexander [197] uses the first setting, with c as the fiber axis. The monoclinic angle γ is obtuse and the shorter unit cell dimension is the a-axis in a right-handed system. This convention is used in most studies that report atomic coordinates (and in Figure 5.17, bottom). These differences are not big problems if there is awareness of the situation.

The diffraction pattern from a vertical fiber has a horizontal equator through the middle of the pattern, along with higher layer lines that are curves if the imaging equipment is flat. There is a vertical line on the pattern that is its meridian. The electron diffraction pattern in Figure 5.17 (right) is reasonably well oriented but only a few diffraction maxima are present. In part, this is because the electron beam tends to severely degrade the small cellulose crystals in cotton. Another approach is to subject homogenized fibrils to acid degradation. The resulting particles can be dried on the inside of a rotating vial, making an oriented film [198].

With diffraction spots falling onto different layer lines, it is much easier to work out the dimensions of the unit cell, the smallest portion of a crystal that can be repeated in all dimensions to construct the entire crystal. The unit cell dimensions are shown in Table 5.1 for the various forms of crystalline cellulose that have recently been studied with high-resolution methods. Figure 5.18 shows the unit cell for cellulose Iβ in a microfibril.

A pattern of intensities that are absent for a given unit cell gives important clues about the space group, the description of the symmetry that applies to that crystalline form. For example, the intensities for the meridional reflections on the first, third, fifth, and higher odd layer lines are absent for the $P2_1$ space group. The number of chains can be inferred from

TABLE 5.1
Unit Cell Dimensions

Allomorph	Å			Degrees		
	a	b	c	α	β	γ
Cellulose Iα	6.717	5.962	10.400	118.08	114.80	80.37
Cellulose Iβ	7.784	8.201	10.380	90.00	90.00	96.50
Cellulose II	8.10	9.03	10.31	90.00	90.00	117.10
Cellulose III$_I$	4.45	7.85	10.31	90.00	90.00	105.10

the dimensions of the unit cell and the density of the sample. Whether there is true symmetry in the cellulose crystallites has always been controversial, with weak odd-order meridional reflections [199] often appearing, thus failing to meet the strict requirement of absence. These odd-order meridional intensities may be artifacts, and in any case, any deviations from symmetry have not been shown to indicate meaningful differences in structure.

When there is only one chain per unit cell, then all of the chains must be parallel. The reducing ends will all be at one end of the crystal and the nonreducing ends at the other. However, if there are two chains per cell, then the chains could be either packed parallel or antiparallel. Antiparallel chains have alternating reducing and nonreducing ends at each end of the crystal. A knowledge of the chain-packing mode is critical to understanding the biosynthetic processes. A much less obvious distinction is whether the parallel chain

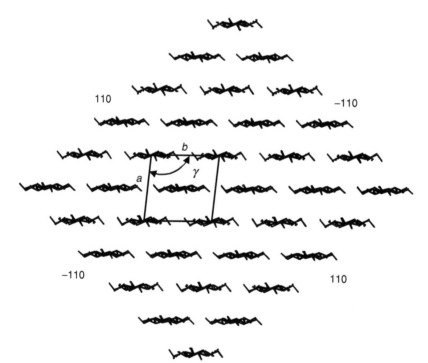

FIGURE 5.18 Cross-sectional view of cellulose Iβ crystallite with 36 molecules. The chain direction is toward the viewer. Also indicated are the *a* and *b* dimensions and the monoclinic angle γ of the unit cell as well as the +110 and −110 faces of the crystallite. A total of two chains pass through the unit cell: the entire central chain and approximately one-fourth of each of the four corner chains.

structures are "parallel-up" or "parallel-down." The convention used to describe the unit cell is important in this definition. A model is said to be parallel-up if the z-coordinate of O5 is greater than that of C5. The difference in packing can be illustrated by the following: the cellulose I model unit cell can be changed from up to down by shifting the adjacent sheets of chains parallel to the b-axis so that the monoclinic angle is changed to acute. Turning the model over then restores the unit cell to the correct, obtuse convention, but the chains are pointed in the opposite direction and have different interchain contacts.

Another issue in structural studies of cellulose is the orientation of the primary alcohol group (Figure 5.11). The O6 atom could be oriented so that it is *gauche* to O5 and *trans* to C4. This position is referred to as *gt*, (*gauche–trans*) or is described as having an O6—C6—C5—O5 torsion angle of approximately 60°. The other orientation found in about equal amounts to *gt* structures in various small-molecule studies is *gauche* to both O5 and C4. This is called the *gg* (*gauche–gauche*) structure (O6—C6—C5—O5 torsion angle ≈ −60°). A third staggered orientation is called *tg* (*trans–gauche*, O6—C6—C5—O5 torsion angle ≈ 180°). The latter is rarely found in small-molecule studies of glucose moieties: two structures show that it can exist, given enough stabilization by hydrogen bonding [200,201].

5.3.3.2 Polymorphs

Cellulose Iα and cellulose Iβ are two different forms that have similar hydrogen bonding systems and crystal packing. Attala and Vander Hart [202] have proposed that all types of native cellulose I are mixtures of the two forms, with Iβ as more important in the higher plants and tunicates and Iα as more important in algae and bacteria. The Iα form is thought to comprise perhaps 10 or 20% of higher plant celluloses such as cotton. After initial studies with NMR, IR, and Raman spectroscopy, the native algal cellulose *Microdictyon tenuius* was shown by electron microscopy to exist simultaneously in two phases, Iα and Iβ, within the same microfibril. Steam annealing converted Iα to Iβ. *Microdictyon tenuius* forms especially large crystallites (300 Å on a side), and the electron diffraction beam was especially narrow [203] With this elegant technique, selected areas of the same microfibril were shown to have both a two-chain unit cell, similar to that long proposed for the common native cellulose, and a new one-chain cell. In some areas, both forms were present, giving the combined diffraction pattern observed with ordinary electron or x-ray diffraction of algal celluloses. The NMR has been used numerous times for delineating the amounts of Iα and Iβ in a given sample. Recent work has elaborated on how the two forms are distributed in the microfibril [204]. A similar intermixture of polymorphs may apply to cotton, except that the fraction of Iα would be much smaller. However, the presence of Iα in higher plant celluloses has been challenged [205,206]. If present in cotton, varying quantities of Iα could lead to variation in the powder diffraction spacings [207] that are routinely ascribed to slightly different unit cell sizes.

The recent cellulose Iα [208] and Iβ [209] crystal structures were obtained with *Glaucocystis nostochinearum* and *Halocynthia roretzi* (tunicin), respectively. They share similar intra- and intermolecular hydrogen bonding systems, with the shifting of the adjacent sheets of chains that contain the intermolecular O—H···O hydrogen bonding systems as the major difference. These sheets are parallel to the crystallographic b axis that is shown in Figure 5.18. In both structures, the crystal packing appears similar when viewed along the chain axes. In both structures, the second sheet is elevated about 2.58 Å relative to the first, and in the Iα structure the third sheet is elevated another 2.58 Å. In Iβ, the third sheet is shifted back down relative to the second sheet, putting it at the same height as the first sheet.

Both cellulose I structures have similar, complex hydrogen bonding systems that contain the O3—H···O5 hydrogen bond that is very typical for β-(1→4)-linked molecules

(Figure 5.19). There are variations in the detailed geometries of those interactions. More importantly, there is substantial variation in O6···O2 intramolecular and O6···O3 intermolecular hydrogen bonds in each polymorph. The hydrogen bonds are described as disordered or having fractional occupancy. Both forms have O6 in the *tg* position, otherwise a rarity for glucose residues. The *tg* orientation of O6 permits a second intrachain, interresidue hydrogen bond, either O2—H···O6′ or O6—H···O2, both of which occur at either various times or various places in the disordered systems. One of the several disordered systems, however, does not have either of the O6···O2 bonds. Similarly, there are weak O6—H···O1 or O2—H···O1 bonds in some but not all of the systems. Perpendicular to the hydrogen-bonded sheets, C—H···O hydrogen bonds have been found, and there are also strong van der Waals

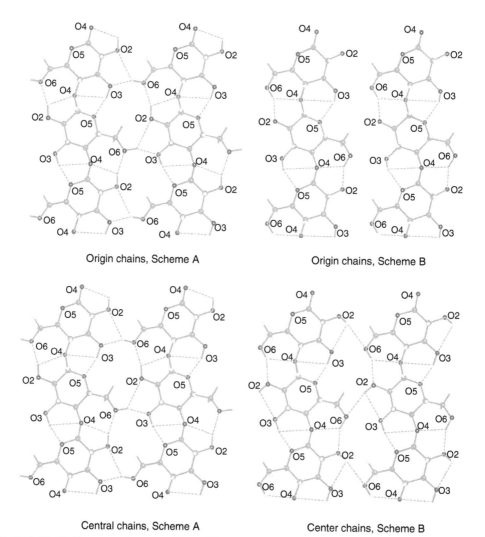

Origin chains, Scheme A   Origin chains, Scheme B

Central chains, Scheme A   Center chains, Scheme B

FIGURE 5.19 Nishiyama et al. reported four different hydrogen bonding schemes that arise in cellulose Iβ because of fractional occupancy of different orientations of the different hydroxyl groups and the slightly different internal geometries of the origin and center chains in the unit cell [209]. Note that the hydroxyl hydrogen atom was not located in the origin **B** scheme. The hydrogen bonds to 04 were not reported by Nishiyama et al. because they appear only with more liberal definition of a hydrogen bond. Even with the more liberal criteria, no interchain hydrogen bonds are found for the origin chains, scheme **B**.

interactions. These attractions do not stabilize the structure as much as the hydrogen bonds within sheets, as indicated by the expansion during heating [210]. Upon heating Iα to 220 or 230°C, the structure converts to Iβ [211]. At this temperature, the intersheet spacing increases enough to allow a sliding shift of the chains to the Iβ form.

Atomic force microscopy (AFM) has been applied to surfaces of *Valonia macrophysa*. The images were compared with surfaces generated from molecular models, and close correspondence was found. The data can be subjected to Fourier transformation to yield a diffraction pattern that can be compared with diffraction patterns made from the model surfaces [212]. The initial images are similar to the lattice images from electron microscopy.

Another AFM study of Valonia cellulose I showed O6 to be in the *gt* orientation [213]. Work with NMR spectrometry has gone further. Difference spectra show that the surfaces have extensively disordered regions as well as *gt* and *gg* O6 orientations [214]. When cellulose I is in water, NMR studies have indicated that its surface hydroxyl groups are also in *gt* positions [215].

The most important alternative crystalline form is cellulose II. This form can result from treatment of cellulose in concentrated alkali, such as 23% NaOH, followed by rinsing in water. This is also the main form that results from crystallization of dissolved cellulose, such as regeneration of rayon. Supercritical water can also effect the transformation [216]. The treatment of cotton in milder alkali, for industrial mercerization, amounts mainly to disruption and decrystallization rather than transformation to crystalline II. Cellulose II can occur as the native state when the normal biosynthesis and subsequent crystallization is disrupted [217–219].

The structure of cellulose II has also been solved by synchrotron x-ray and neutron diffraction. Although there was speculation that mercerized cellulose II might be parallel while regenerated cellulose II was antiparallel, the same antiparallel packing was found for regenerated Fortisan, mercerized flax [220] and mercerized ramie [221]. The O6 atoms are in the more widely observed *gt* position although there is fractional occupancy of the *tg* orientation, more so in regenerated cellulose. Although there is a long-range attraction from O3—H to O6 between rings on the same chain, the modulus for chain extension is considerably lower for cellulose II, which lacks the strong O6···O2 hydrogen bonds found for cellulose I [194]. Unlike cellulose I, there is a three-dimensional intermolecular O—H···O hydrogen bonding system but there is no C—H···O hydrogen bonding.

The cellulose oligomers, beginning with methyl cellotrioside, yield powder diffraction patterns that are very similar to those of cellulose II. The NMR studies of the cellulose oligomers further establish the extensive analogy between cellotetraose and cellulose II. Work by both Gessler et al. [222] and Raymond et al. [223] has shown that the O6 atoms in cellotetraose and methyl cellotrioside [224] all take the *gt* position, consistent with the diffraction and NMR results for cellulose II. Because the chains in the methyl cellotrioside and cellotetraose are antiparallel, this work adds support to the above results on cellulose II. On the other hand, molecules in crystalline α-lactose, a related disaccharide, have parallel packing [225].

Cellulose III results from treatment in amines or liquid ammonia. The finding of only six peaks in the ^{13}C NMR data for cellulose III$_I$ (the form of cellulose III made starting with cellulose I) was a sensitive indicator of a single glucose residue in the asymmetric unit [226], and the NMR data also defined the O6 position. Besides the implications from a single-chain unit cell [227], parallel packing in cellulose III$_I$ has recently been confirmed by a high-resolution study that located all the atoms, including hydrogen [228]. The sample was *Cladophora* seaweed cellulose, treated with supercritical ammonia, and the *gt* O6 positions allow an extensive cooperative hydrogen bonding network. Except for its one-chain unit cell, and therefore parallel chains, the packing and hydrogen bonding are reasonably similar to that of cellulose II.

Cellulose IV results from treatments at high temperatures, such as in glycerol at 260°C. The early cell wall material from some cells was reported to be cellulose IV [229]. The latest information is that cellulose IV made from cellulose I is actually a less crystalline cellulose I [226]. The NMR was instrumental in developing that idea. This is consistent with the finding of cellulose IV in the primary cell wall [229], but more developments are expected on this subject. (I just heard a paper at the ACS meeting in which the NMR work showed that the cellulose IV sample used in the x-ray study may not have been properly prepared.)

Cellulose Iβ and cellulose II have monoclinic, two-chain unit cells. Cellulose III$_I$ has a monoclinic one-chain cell [227], and the one-chain unit cell of Iα is triclinic with no 2_1 symmetry. Still, all of the chain shapes are very similar to each other. It had been speculated that cellulose chain linkage geometries would alternate between the quite different linkages found in crystalline β-cellobiose and in methyl β-cellobioside [196]. That idea is now obsolete. Such a departure from symmetry would be far greater than indicated by the above high-resolution studies. When molecules from the high-resolution structures for all of the polymorphs are superimposed, differences in their backbone structures are barely visible.

Solid-state cellulose can also be noncrystalline, sometimes called amorphous. Intermediate situations are also likely to be important but not well characterized. One example, "nematic ordered cellulose" has been described [230]. In most treatments that produce amorphous cellulose, the whole fiber is severely degraded. For example, decrystallization can be effected by ball milling, which leaves the cellulose as a fine dust. In this case, some crystalline structure can be recreated by placing the sample in a humid environment. Another approach uses phosphoric acid, which can dissolve the cellulose. Precipitation by dilution with water results in a material with very little crystallinity. There is some chance that the chain may adopt a different shape (a collapsed, sixfold helix) after phosphoric acid treatment. This was concluded because the cellulose stains blue with iodine (see Figure 5.12), similar to the sixfold amylose helix in the starch–iodine complex.

One of the most special aspects of cellulose polymorphy is the transformation from I to II. The conversion of the parallel-packed cellulose I structures to an antiparallel cellulose II structure is interesting because it can occur without loss of the fibrous form. This transformation is widely thought to be irreversible, although there are several reports [231–233] of regenerated cellulose I. The observation that there are two different forms of cellulose III and of IV is also remarkable. The two subforms of each allomorph have essentially identical lattice dimensions and at least similar equatorial intensities. Other intensities are different, particularly the meridional intensities, depending on whether the structures were prepared initially from cellulose I or II. The formation of the III and IV structures is reversible and the preceding polymorph (I or II) results.

Work in the 1970s proposed the apparently correct explanation for the above reversible and irreversible transformations, i.e., that the chains in cellulose I are packed parallel and in cellulose II are antiparallel. However, there were problems with the work that left room for genuine doubts. These doubts were gradually erased by many separate studies. Now, it appears that conversion from I to II with the retention of the fibrous form is unique to cellulose in secondary cell walls in which the adjacent crystallites are tightly packed and constrained by a primary wall [234,235]. Extracellular celluloses and loosely packed, parenchymal or primary wall celluloses do not retain the fibrous form after treatment in strong NaOH. Instead, they contract into lumps that exhibit cellulose II diffraction patterns. These lumps could contain folds that would account for the antiparallel chains; a model fold is shown in Figure 5.12. The explanation for the ability to convert from parallel to antiparallel chain packing with retention of fibrous form is that each fibril is composed of parallel molecules in the original fibers, but that the fibrils themselves are antiparallel. During treatment with NaOH, the fibers swell and the molecules from the adjacent antiparallel fibrils interdigitate. This allows the formation of the lower-energy crystalline form.

Besides the cellulose structures I–IV and their subclasses, cellulose forms a variety of crystalline complexes. Soda celluloses were mentioned above, and there is an extensive array of complexes with amines [236]. Soda cellulose IV [237] is actually a hydrate of cellulose and contains no sodium (historically, cellulose hydrate meant cellulose II, which is now known to contain no water!). Many cellulose derivatives such as the nitrate (see above) and the triacetate [238] also give diffraction patterns. The most recent analysis of triacetate I shows a single-chain unit cell [239].

5.3.4 Electron Microscopy and Lattice Imaging

Work with electron microscopes showed that there is preferential enzymatic activity at only one end of the native microfibrils. This indicates that the reducing ends are all at one end of the microfibril and thus the chains are parallel, not antiparallel [240]. Electron microscopy and diffraction work on algal and bacterial cellulose confirmed the parallel-up nature of the chain orientation in the unit cell and the addition of new glucose residues to the cellulose chain at the nonreducing end [241]. Similar attempts with ramie fibers were not successful.

Lattice images of cellulose can be obtained from cellulose samples in the electron microscope and subjected to the same Fourier transformations as AFM images. Both of these techniques confirm the idea that cellulose chains are very extended in crystalline microfibrils and emphatically do not undergo folding within linear microfibrilar structures, as had been proposed by some authors.

Lattice images of algal, bacterial, and ramie cellulose have been obtained. These images show the individual molecular chains and the sizes of microfibrils, which vary in size and shape according to the source of cellulose [242,243]. There is also some variation within a given source. For example, microfibrils of Valonia ranged from 150 to 250 Å (15 to 25 nm). No evidence of elementary fibrils was seen.

5.3.5 Small Angle X-Ray Scattering

Although conventional x-ray diffraction equipment does not permit large (>100 Å) structures to be studied, a special apparatus can be configured to detect the behavior at very small scattering angles. Such devices are often used to study synthetic polymers. One such experiment on various cellulose samples was able to detect the pores of cotton fibers [244]. The results for normal cotton depended on a novel fractal analysis of the data rather than on the classical Guinier and Porod analyses. The void volume fraction ranges from 0.7 to 3.4% in the cotton samples and was 17% in Valonia. Dewaxing, scouring, and bleaching increased the void volume, where NaOH mercerization and ammonia treatments decreased not only the packing efficiency but also the void volume. In hydrocellulose II, an acid-hydrolyzed cotton cellulose II, the average pore size was 85 Å and the specific inner surface area was 15.3 m^2/cm^3.

5.3.6 Diffraction Studies of Crystallite Size

Cellulose powders can be created by cutting fibers into small particles, perhaps with a Wiley mill (Arthur H. Thomas Company, Swedesboro, New Jersey). On a laboratory x-ray system, powder diffraction patterns take 30 min. The positions of the peaks indicate the polymorphic form (I–IV); the powder diffraction pattern is often used as a fingerprint for comparison with the known pattern for a given crystalline form [207]. The breadth of the peaks is related to the extent of crystallinity (Figure 5.17, bottom). Using the Scherrer formula [245,246] and assuming no other distortions, the crystallite size can be calculated. Values for cotton perpendicular to the molecular axis are around 40 Å. That corresponds to a 6×6 array of cellulose chains (Figure 5.19). (Note that 20 of these 36 chains constitute the surfaces of the crystallite.) The crystallinity index (CI) can be simply calculated by comparing the minimum intensity just before the largest peak with the peak height [247]. By this simple method, the crystallinities for various varieties of cotton are in the range of 80%. Other methods for measuring crystallinity are based on different physical phenomena and the absolute results can be quite different. The rank ordering is usually similar, however.

5.3.7 CRYSTALLITE ORIENTATION BY DIFFRACTION

Another important aspect of cotton structure, the crystallite orientation, can be determined with diffraction analysis of bundled fibers [248]. The degree of arcing of the spots on fiber patterns (e.g., Figure 5.17, left) indicates the extent to which the long axes of the crystallites are not completely parallel to the fiber axis. The deviation from perfect alignment is an important indicator of fiber strength (see Section 7.2). In the case of cotton fiber bundles, the spots on the diffraction pattern are large arcs, resulting from the spiral arrangement of the microfibrils, with reversals (see Section 5.5). The molecular axes in cotton fibers are not parallel to the fiber axis. Instead, different samples have average deviations from about 25 to 45°. Cotton is considerably weaker than ramie or linen, both of which have much smaller arcs. If a slurry of cotton fibers in water is cut into small particles in a tissue homogenizer, the spiral structure is broken up and the electron diffraction pattern of the individual particles (Figure 5.17, right) is similar to that for a ramie fiber bundle [249]. That work showed that the arc lengths on an electron diffraction pattern from cut up ramie particles are even shorter.

5.4 CRYSTAL AND MICROFIBRILLAR STRUCTURE BY CHEMICAL METHODS

Much information about the structure of cellulose can be gleaned from data using sorption and chemical reactivity techniques. Information is provided on the accessibility and reactivity of the cellulose sample. An important aspect of such data is the fact that it must be interpreted as accessibility and reactivity to the specific agent and the test conditions used. This subject has recently been reviewed [250,251].

5.4.1 SORPTION

Deuterium, moisture, iodine, and bromine sorption have been utilized for investigating the supramolecular structure of cotton and mercerized cotton. The methods have been described elsewhere [251]. Average ordered fractions are given in Table 5.2.

In some instances, nonaccessibility, or the so-called average ordered fraction, is measured rather than crystallinity. Values vary depending on the size of the probe molecule and its ability to penetrate and be adsorbed in all the disordered regions. It will be noted that the average ordered fraction is relatively close for the deuteration and moisture regain methods. In addition, the average ordered fraction is decreased about 25% by mercerization.

Deuteration of the accessible hydroxyl groups is accomplished with saturated deuterium oxide vapor at room temperature. The extent of deuteration and therefore accessibility can be estimated gravimetrically by infrared spectroscopy. Accessibility rather than crystallinity is measured because deuteration of the hydroxyl groups on crystallite surfaces can also occur.

Water vapor at room temperature will not penetrate well-defined crystallites but will be adsorbed in the amorphous regions. Consequently, moisture sorption measured gravimetrically at a given relative vapor pressure and temperature has been used to determine order in cellulosic materials. In the case of Valentine [252] and Jeffries [253], the fraction of ordered material was obtained by correlating moisture sorption with values obtained by the deuterium

TABLE 5.2
Average Ordered Fraction (AVF)[a,b] and Crystallinity (XAL)[b] of Cotton and Mercerized Cotton Determined by Sorption

Technique	Cotton		Mercerized Cotton		Ref.
	AVF	XAL	AVF	XAL	
Deuteration	0.58	—	0.41	—	[251]
Moisture Regain					
Valentine	0.615	—	0.45[c]	—	[252]
Hailwood-Horrobin	0.67	—	0.50	—	[251]
Jeffries	0.58	—	0.465	—	[253]
Zeronian	0.63	0.91	0.48	0.79	[261]
Iodine	0.87	—	0.68	—	[251]
Bromine	0.46–0.80[d]	—	—	—	[255]

[a]AVF can be considered the nonaccessible fraction of the sample since the measurements consider crystallite surfaces to be part of the low-order regions.
[b]Expressed as a fraction.
[c]Calculated from Valentine's relation [251] using a value of 1.43 for the sorption ration of mercerized cotton [260].
[d]Depending on cotton variety.

oxide method. Hailwood and Horrobin [254] developed an equation for water sorption of cellulose based on a solution theory that allowed the calculation of the fraction of the sample inaccessible to water.

Lewin and coworkers [255–260] developed an accessibility system based on equilibrium sorption of bromine, from its water solution at pH below 2 and at room temperature, on the glycosidic oxygens of the cellulose. The size of the bromine molecule, its simple structure, hydrophobicity, nonswelling, and very slow reactivity with cellulose in acidic solutions, contribute to the accuracy and reproducibility of the data obtained. The cellulose (10 g/l) is suspended in aqueous bromine solutions of 0.01–0.02 mol/l for 1–3 h, depending on the nature of the cellulose, to reach sorption equilibrium. The diffusion coefficients of bromine in cotton and rayon are 4.6 and 0.37×10^{-9} cm^2/min, respectively. The sorption was found to strictly obey the Langmuir isotherm, which enables the calculation of the accessibility of the cellulose as follows:

At equilibrium,

$$C_B = KC_f(C_{B(s)} - C_B) \tag{5.1}$$

where C_B is the concentration of bromine in mol/kg of cellulose (e.g., the number of occupied sites), $C_{B(s)}$ the saturation concentration of bromine in mol/kg, $C_{B(s)} - C_B$ the number of unoccupied sites; C_f the equilibrium concentration of bromine in mol/l of solution; and K the equilibrium constant.

If we put the accessibility $A = 100/n$ and $n = m/C_{B(s)}$, where $1/n$ is the number of anhydroglucose units (AGUs) available for sorption, assuming that one AGU is accessible to one bromine molecule, and $n = 1000/162$, e.g., the number of mole AGUs in 1 kg of cellulose, we obtain:

$$\left(\frac{K}{n}\right)\left(\frac{m}{C_B}\right) - K = \frac{1}{C_f} \tag{5.2}$$

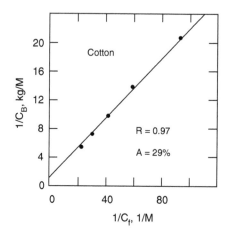

FIGURE 5.20 Langmuir sorption isotherm of Br_2 on cotton fibers.

The value of n is calculated from the extrapolation of the straight line obtained by plotting m/C_B against $1/C_f$ to the value of $1/C_f = 0$ (see Figure 5.20).

Straight-line relationships were obtained between the accessibilities by the bromine method and the IR crystallinity indices and the wide angle x-ray scattering (WAXS) indices for 16 native and regenerated celluloses with accessibilities ranging from 70 to 6% (See Table 5.3 and Figure 5.21 and Figure 5.22).

The accessibilities determined by the bromine method involve only the noncrystalline, less-ordered regions (LORs) and do not include the surfaces of the crystalline regions. The glycosidic oxygen on these surfaces is buried half a molecule deep within the crystallite [260] and is inaccessible both to H_3O^+ which is responsible for hydrolysis, as well as to bromine molecules. Unlike the glycosidic oxygen, the hydroxyl groups protrude from the crystallite surfaces and are, therefore, accessible to deuteration and moisture absorption, which yield similar accessibilities [260]. Substitution reactions similarly occur on the hydroxyl groups.

Zeronian et al. [261] hypothesized that if microcrystalline cellulose is prepared that is a facsimile of the crystalline regions present in the fiber then the fraction of amorphous material (F) of the fiber can be obtained from the relation

$$F = \frac{(M_s - M_c)}{(M_a - M_c)} \tag{5.3}$$

where M_s, M_c, and M_a are the monomolecular moisture regains of the sample, its microcrystalline counterpart, and amorphous cellulose, respectively. F is a measure of the disordered cellulose that does not include crystalline surfaces. Accessibility (A_s) is given by

$$A_s = \frac{M_s}{M_a} \tag{5.4}$$

The fraction of crystalline material (X) is given by

$$X = 1 - F \tag{5.5}$$

The values of X obtained by Zeronian et al. [261] for cotton and mercerized cotton agree well with the values for fraction of ordered material obtained by acid hydrolysis.

TABLE 5.3
Accessibilities and Crystallinity Indices (CI) of the Cellulose Samples

Sample	$n = \dfrac{1}{C_{B(s)}} \times m^a$	S^a	R^b	Accessibility by Br Method, $A = 100/n$	CI by X-ray	CI by IR $\dfrac{1429\,cm^{-1}}{893\,cm^{-1}}$	CI by IR $\dfrac{1372\,cm^{-1}}{2900\,cm^{-1}}$
1. Pay master cotton fibers	4.54	0.14	0.97	22	0.785	2.56	0.59
2. Amsark cotton fibers	3.7	0.12	0.97	27	0.740	2.3	0.55
3. Pima cotton fibers	5.0	0.10	0.96	20	0.780	2.44	0.66
4. Acala 4–420 cotton fibers	4.34	0.06	0.98	23	0.750	2.5	0.64
5. Bradley cotton fibers	3.4	0.11	0.97	29	0.740	2.2	0.57
6. Moores cotton fibers	1.85	0.11	0.97	54		1.6	0.45
7. Pima cotton fibers	3.1	0.16	0.98	32	0.700	2.2	0.51
8. Cross-linked Pima cotton fabric	9.09	0.07	0.92	11	0.830	2.74	0.65
9. Hydrolyzed Pima cotton fabric	7.14	0.11	0.97	14	0.860	2.72	0.6
10. Ramie fabric	5	0.11	0.98	20	0.800	2.46	0.59
11. Hydrolyzed ramie fabric	15.2	0.11	0.98	6.5	0.880	3.07	
12. Evlan fibers	3.7	0.10	0.98	27	0.720	2.26	0.53
13. High modulus rayon fibers	4.46	0.10	0.97	22.4	0.730	2.6	0.54
14. Vincel 28 fibers	2.11	0.11	0.97	47	0.630	1.76	0.48
15. Vincel 66 fibers	3.25	0.10	0.97	30.7	0.740	2.35	0.53
16. Cellulose triacetate	1.42	0.037	0.97	70		1.07	0.37

[a]The plot of $1/C_B$ against $1/C_f$ is a straight light with the slope $S = n/Km$.
From the intercept $1/C_{B(s)}$, $n = (1/C_{B(s)}) \times m$ is calculated.
[b]R = degree of correlation.
Source: From Lewin, M., Guttman, H., and Saar, N., *Appl. Polym. Symp.*, 28, 791, 1976.

5.4.2 ACID HYDROLYSIS

Acid hydrolysis is usually carried out with mineral acids at elevated temperatures. A portion of the cellulose reacts much faster than the remainder under these conditions. It is believed that the initial reaction occurs in the disordered regions and later extends to the ordered regions. The chain cleavage occurs and the products are glucose, soluble oligosaccharides, and an undissolved residue designated hydrocellulose. The weight of hydrocellulose is plotted

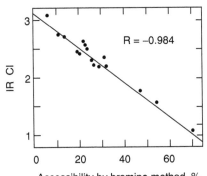

FIGURE 5.21 Infrared (IR) crystallinity index (CI) 1429 cm^{-1} versus accessibility by the Br$_2$ method for different cellulose fiber.

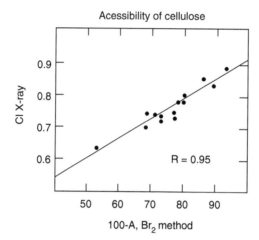

FIGURE 5.22 X-ray crystallinity index (CI) versus accessibility by the Br$_2$ method.

against time [262] (Figure 5.23). Resolution of the plot into two components and extrapolation of the data at the slower rate to zero time gives an estimate of the ordered fraction. The sum of the initial percentages of crystalline and amorphous cellulose is always less than 100%. This may be due to incomplete analysis of the weight-loss data. Estimates of the degree of order by this technique may be high. Here the chemical agent is quite small and the acidic medium might disrupt the ordered regions that would permit the reaction to continue.

5.4.3 FORMYLATION

The formylation method is based on the determination of the ratio of the extent of esterification of cellulose by formic acid after a given time interval to that of soluble starch for the same

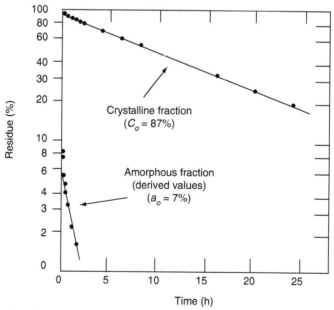

FIGURE 5.23 Relationship between residue weight and time of hydrolysis at 100°C with 6 M HCl for cotton linters. (From Nelson, M.L., *J. Polym. Sci.*, 43, 351, 1960.)

time interval. It is assumed that the starch is fully accessible to the reagent; thus a measure of the accessible fraction of the cellulose can be calculated [263,264]. By extrapolating the plot of this ratio against time to zero time, the initial accessible fraction of the sample can be determined [263]. The complement of this value is the ordered fraction. Other workers had arbitrarily measured accessibilities after 16 h of esterification [264–266]. However, much closer agreement with acid hydrolysis values is obtained if the initial ordered fraction is obtained. Nickerson's estimates of the initial nonaccessible fractions of cotton, mercerized cotton, and viscose rayon are 0.91–0.92, 0.85–0.86, and 0.70–0.72, respectively [263]. The advantages of this method are that the reaction is autocatalytic; the formic acid molecule is small, polar, and water miscible; the reagent is a relatively strong swelling agent for cellulose but does not penetrate the ordered regions [263]. A disadvantage is some chain scissions may occur [264] and result in crystallization.

5.4.4 PERIODATE OXIDATION

This method is based on the preferential oxidation of the disordered regions by sodium metaperiodate [266,267]. Conditions are selected so that the reaction is confined as far as possible to the Malaprade course resulting in the formation of 2,3-dialdehyde units. The course of the reaction is followed by measuring the oxidant consumption from the amount of periodate consumed. From plots of log oxidant consumption against time, a measure of the fraction of ordered material can be calculated analogous to that of the acid hydrolysis method.

5.4.5 CHEMICAL MICROSTRUCTURAL ANALYSIS

Chemical microstructural analysis (CMA) method is based on reactivity of the cellulosic hydroxyl groups with diethylaminoethyl chloride under very mild basic conditions, which, to the best of our knowledge, does not further disrupt ordered regions. The cotton cellulose is reacted with diethylaminoethyl (DEAE) chloride as shown in Figure 5.24.

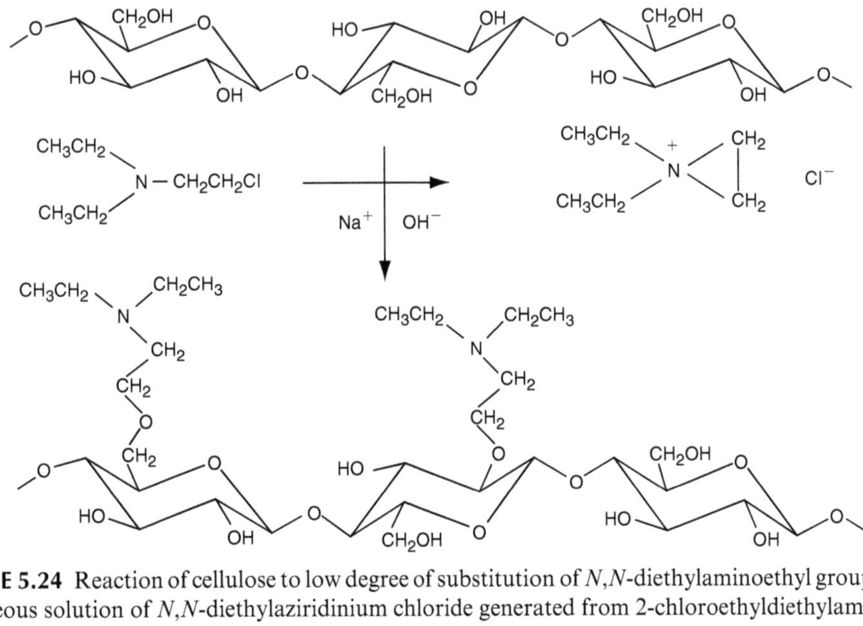

FIGURE 5.24 Reaction of cellulose to low degree of substitution of *N,N*-diethylaminoethyl groups occurs in aqueous solution of *N,N*-diethylaziridinium chloride generated from 2-chloroethyldiethylamine.

TABLE 5.4
Fraction of Total Reactivity

Hydroxyl	Fraction
OH-2	0.655
OH-3	0.059
OH-6	0.286

The DEAE cellulose is then hydrolyzed to substituted glucoses and glucose, which are silylated. The relative quantities of OH-2, OH-3, and OH-6 DEAE glucose are determined via gas chromatography. The fraction of total reactivity under these conditions for each of these hydroxyls is given in Table 5.4.

Note that under these nondisruptive, basic conditions the most reactive hydroxyl is OH-2. In this sample, the fraction of total reactivity of OH-2 is 11 times that of OH-3 and more than double that of the primary hydroxyl OH-6. This is quite different from relative reactivity observed for soluble glucans in which OH-6 is usually the most reactive hydroxyl. The CMA technique has been discussed in detail elsewhere [250,251]. Because OH-2 is the most reactive in this reaction, data is reported relative to this hydroxyl. Representative results for several cottons are given in Table 5.5.

The relative availability of each hydroxyl group of totally amorphous cellulose is 1.00 by definition. Mercerization converts the native cellulose I to cellulose II. Hydrocellulose is the unreacted cellulose after acid hydrolysis, which removed the more disordered regions as discussed above. The native cellulose for which the data is presented is cotton that has been dried. Drying alters the relative availability of the OH groups [268]. In agreement with the predictions (Figure 5.13) based on the unit cell structure, the OH-3 to OH-2 ratio has been shown to approach a limiting value of zero in a never-dried cotton. It is assumed that this represents almost perfect crystallinity. This value increases upon desiccation of the fiber on boll opening [268] (Figure 5.25).

This is the first introduction of stress and disorder into the fiber. Greige mill processing further increases this microstructural disorder as evidenced by increasing availability of the OH-3 as shown in Figure 5.26 [269].

5.4.6 SUMMARY OF AVERAGE ORDERED FRACTION VALUES DETERMINED BY CHEMICAL METHODS

The values for the average ordered fractions in cotton and mercerized cottons determined by the different chemical methods are summarized in Table 5.6.

TABLE 5.5
Relative Availabilities of Hydroxyls

Cotton	OH-3/OH-2	OH-6/OH-2
Native	0.27	0.82
Mercerized	0.79	0.86
Hydrocellulose	0.23	0.74
Amorphous	1.00	1.00

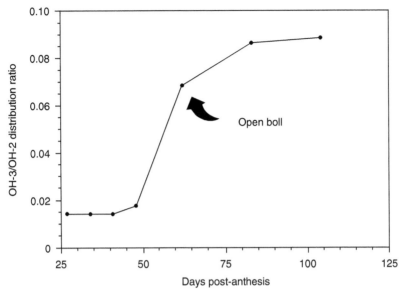

FIGURE 5.25 OH-3/OH-2 distribution ratios for cotton fibers as a function of days post-anthesis (DPA).

These data for cotton and mercerized cotton are comparable to those determined by the sorption techniques (Table 5.2), by x-ray diffraction (0.73 and 0.51, respectively), and density (0.64 and 0.36, respectively) [250].

5.5 FIBER STRUCTURE

5.5.1 MORPHOLOGY

Cotton is a unique textile fiber because of the interrelationships of its subunits. From its multicomponent primary wall, through the pure cellulose secondary wall to the lumen, the organization of subfiber units provides the fiber with characteristic properties that make it a

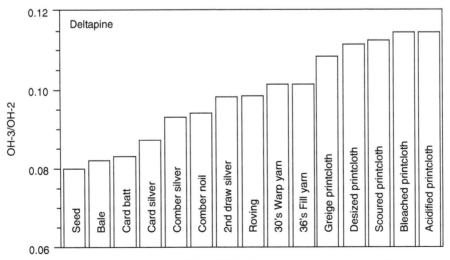

FIGURE 5.26 Effect of processing on the OH-3/OH-2 distribution ratio in cotton fibers.

TABLE 5.6
Average Ordered Fraction in Cotton and Mercerized Cotton
Determined by Chemical Methods

Method	Cotton	Mercerized Cotton
Acid hydrolysis	0.90	0.80
Formylation	0.79	0.65
Periodate oxidation	0.92	0.90
Chemical microstructural analysis[a]	0.73	0.29

[a]Based on availability of 3-OH.

processible, strong, and comfortable textile fiber. The outer skin of the fiber, the cell wall (cuticle-primary wall), is composed of an inner network of microfibrils randomly organized within a mixture of waxes, pectins, proteins, and other noncellulosic materials [270]. In the dried fiber, the microfibrils are present in a network of crisscrossing, threadlike strands that encase the entire inner body of the fiber. The noncellulosic components of the cell wall give the fiber surface a nonfibrillar appearance, and provide both a hydrophobic protection in the environment, and a lubricated surface for processing (Figure 5.27 and Figure 5.28).

It is the waxy component of the primary wall that must be partially removed to allow the finishing and dyeing chemicals to access the body of the fiber. Inside the primary wall, a thin layer, called the winding layer (Figure 5.29), consists of bands of helical microfibrils that are laid down in a lacy network, which has been associated both with the primary wall [271] and with the secondary wall [99,272].

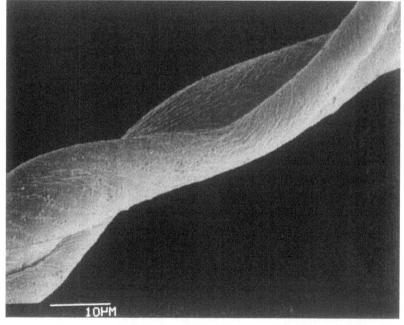

FIGURE 5.27 Typical dimensional structure variation in cotton fiber after drying (scanning electron microscope, SEM).

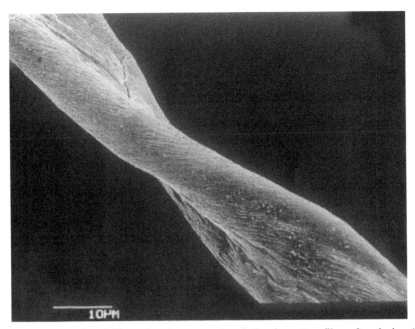

FIGURE 5.28 Less convoluted dimensional structural variation in cotton fiber after drying (SEM).

During fragmentation of the fiber, the winding layer often separates with the primary wall; however, it is believed that the winding layer cellulose is deposited at the time of decreasing elongation when secondary wall synthesis begins, and therefore may be more closely connected chemically to the secondary wall. The intermeshed fibrillar network of the primary wall and the woven mat of fibers of this winding layer just beneath it provide a dynamic casing that allows limited swelling within the secondary wall, which has its

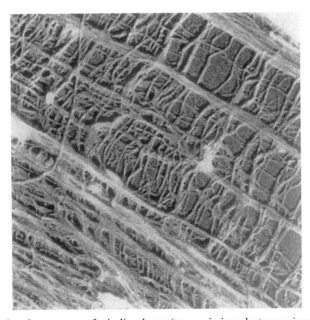

FIGURE 5.29 Cross-hatch structure of winding layer (transmission electron microscope, TEM).

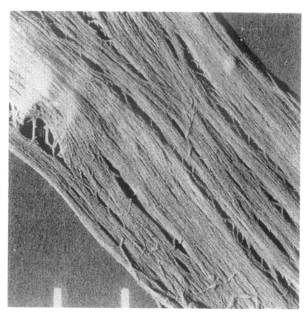

FIGURE 5.30 Parallel microfibrils of secondary wall (TEM).

microfibrils oriented more along the fiber axis. The casing protects the secondary wall fibrils from lateral separation forces during swelling. As long as the primary wall-winding layer is intact, the inner portion of the fiber is less accessible to damage. The main body of the fiber, the secondary wall, consists of layers of nearly parallel fibrils laid down concentrically in a spiral formation (Figure 5.30).

Secondary wall fibrils closer to the primary wall lie at an approximate 45° angle to the fiber axis (Figure 5.31), while this orientation becomes aligned more closely with the fibrillar axis as the fiber core, or lumen, is approached (Figure 5.32).

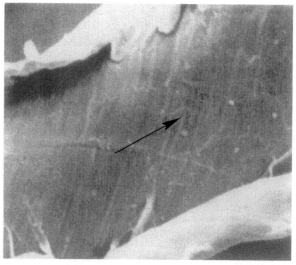

FIGURE 5.31 Fiber with peeled primary wall revealing underlying secondary fibrils at 45° to fiber axis (SEM). Arrow indicates fiber axis.

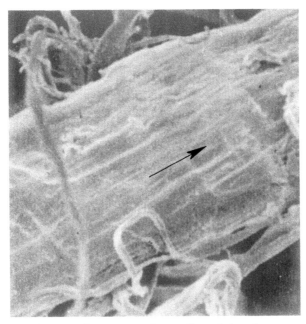

FIGURE 5.32 Fiber with primary and outer secondary walls peeled to show secondary wall fibrils lying almost parallel to fiber axis (SEM). Arrow indicates fiber axis.

The direction of the spiral around the axis of the fiber reverses at random intervals along the length of the fiber. These fibrillar directional changes are called reversals, and can be detected by following the direction of wrinkles on the fiber surface (Figure 5.33).

Reversals represent zones of variations in breaking strength. The areas immediately adjacent to either side of a reversal are more likely to break under stress than are other

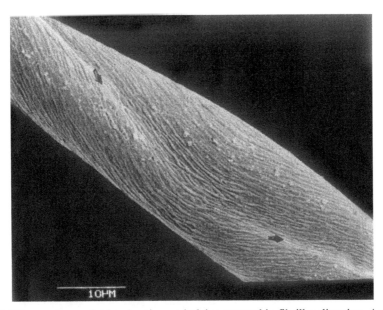

FIGURE 5.33 Fiber surface wrinkles showing underlying reversal in fibrillar direction. Arrows indicate direction of fibril wrap (SEM).

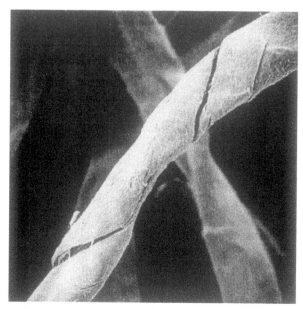

FIGURE 5.34 Fiber showing cracks developed on either side of a reversal (SEM).

fiber areas [65,273]. Figure 5.34 shows cracking of a fiber damaged by fungus on each side of a reversal, as indicated by wrinkles in the fiber's primary wall.

The densely packed fibril layers of the secondary wall are considered to be pure cellulose. The thickness of this wall, from primary wall to lumen, is closely related to gravimetric fineness (mass per unit length). Fibers with no secondary wall development exhibit no individual fiber integrity and can exist only in clumps [274]. Figure 5.35 illustrates cross section of a bundle of fibers with primary wall, but no secondary wall. Development of the secondary wall provides the fiber with rigidity and body.

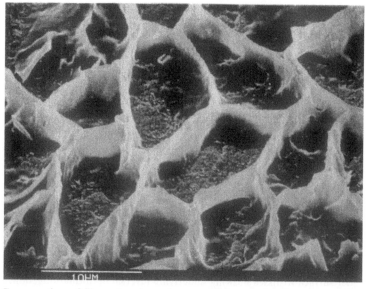

FIGURE 5.35 Cross-section of fibers showing primary wall but no developed secondary wall (SEM).

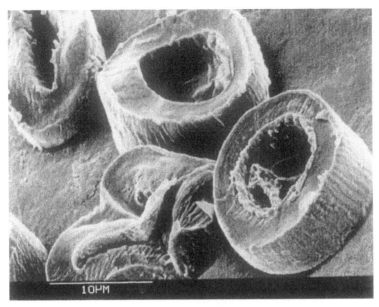

FIGURE 5.36 Cross-section of fibers showing partially developed primary walls, immature fibers (SEM).

Figure 5.36 shows intermediary secondary wall development (immature fibers) in cross section, while Figure 5.37 illustrates cross sections of mature fibers.

These sections were harvested live and processed in the wet state so they are more rounded, and have not assumed the characteristic Kidney-bean-shaped fiber cross-sectional shape of the dried fiber. Fibers with thinner secondary walls are known as immature, while those with walls at or approaching their maximum thickness are called mature. Thus, maturity is a relative term that is difficult to measure objectively (see Section 7.1). Secondary wall thickness is directly

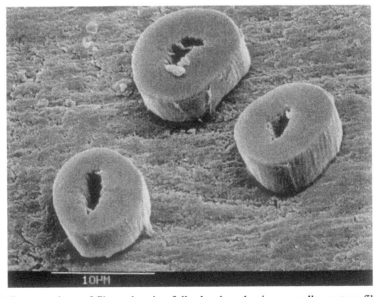

FIGURE 5.37 Cross-sections of fibers showing fully developed primary walls, mature fibers (SEM).

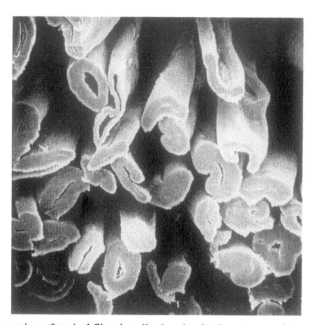

FIGURE 5.38 Cross-section of typical fiber bundle showing both mature and immature fibers (SEM).

related to fiber properties such as strength, dyeability, and reactivity. Figure 5.38 is a cross section of a typical fiber bundle, showing a mixture of both mature and immature fibers.

When the fiber is dried, no differentiation between successive layers of secondary wall fibers can be distinguished. However, when fibers are swelled and viewed at higher magnifications in cross section, lacey, layered patterns become apparent [275]. Figure 5.39 illustrates layering that occurs when fibers wet with water or other liquids such as lower alcohols, ethylene glycol, or glycerin are embedded by polymerization of methyl and butyl methacrylates.

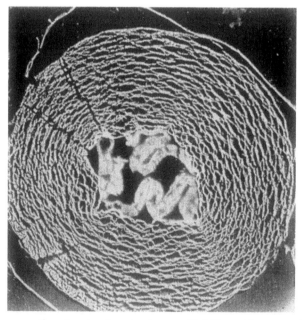

FIGURE 5.39 Thin cross-section showing layering in a swelled fiber at low magnification (TEM).

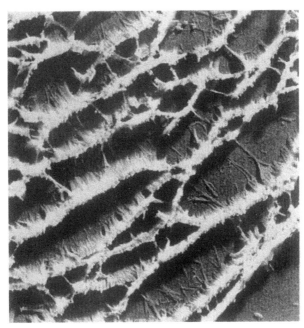

FIGURE 5.40 Portion of thin cross-section showing fibrils in layers of a swelled fiber at higher magnification than Figure 5.39 (TEM).

The open spaces within the fiber represent sites of entry for the liquids, and thus accessibility of the fiber to the liquid [276]. Figure 5.40 shows this layering at higher magnification, revealing the fibrils that compose the layers of the secondary wall.

The lamellar pattern is believed to be due to the differential rates of fibrillar formation during daily growth cycles [277,278]. Differential compaction of the fibrils provides variations in fibrillar bonding, and determines accessibility or permeability into the inner areas of the fiber. The lumen is the central opening in the fiber that spans its length from base nearly to the tip. It contains the dried residues of cell protoplasm, the only source of noncellulose materials in the fiber other than those in the primary wall. A thin cell wall (lumen wall) provides the inner cell boundary. The lumen opening occupies about 5% of the cross-sectional area of the mature fiber.

The live, hydrated fiber exists in a tubular configuration that conforms to available space within the boll locule. When the boll opens, removal of water causes the internal layers of the fiber to twist and collapse, and the primary wall, which is less able to shrink because of its network structure, wrinkles and molds to the underlying fiber layers, producing folds and convolutions (twists), and compression marks (Figure 5.41). Fibers often collapse in a nonuniform elliptical pattern, whose cross section has a convex and a concave side (see Figure 5.27 and Figure 5.38).

This pattern is more pronounced in fibers of low maturity. Even in mature fibers, the lumen cross section assumes an elongated shape on drying, thus giving the fiber cross section a long and a short axes. This asymmetric structure indicates that there may be differences in fibrillar packing densities around the perimeter of the fiber. Such zones would present different areas of accessibility in the fiber. It is not known whether these zones of variations in fibrillar density are due to inherent differences in fibrillar structure at different areas of the cross section, or whether physical forces during drying compress the structure in some areas and expand it in others [279,280]. Dried fibers with relatively thick secondary walls produce

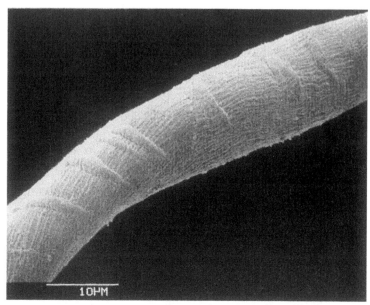

FIGURE 5.41 Surface of a cotton fiber showing compression marks due to removal of water on drying (SEM).

the thick, bean-shaped cross section usually associated with the structure of the cotton fiber. This drying and shrinking process produces the nonuniform, convoluted cotton textile fiber, which, although it has its basis in the living biological fiber, is structurally quite different in the dried state than in the living, hydrated state.

5.5.2 PORE STRUCTURE

The cotton fiber is hydrophilic and porous. Upon immersion in liquid water, the cotton fiber swells and its internal pores fill with water. Pure cotton holds a substantial percentage of its dry weight in water under conditions of centrifugation. The amount of liquid water held depends upon the severity of the centrifugation used in testing. This is approximately 30% for the water of imbibition [281] or 50% for the water retention value [282]. Centrifugation conditions are less severe in the latter case.

Pores accessible to water molecules are not necessarily accessible to chemical agents. Chemical modification is required to impart many desired properties to cotton fabric. These include color, permanent press, flame resistance, soil release, and antimicrobial properties to name a few. Thus, a knowledge of cotton's accessibility under water-swollen conditions to dyes and other chemical agents of various sizes is required for better control of the various chemical treatments applied to cotton textiles.

The accessibilities of cotton fibers have been measured by solute exclusion. A simplified mechanism is shown in Figure 5.42.

This subject has recently been reviewed [283]. Both static techniques [284], glass column chromatography [285] and liquid chromatography [286,287], have been used. Series of water-soluble molecules of increasing size are used as molecular probes or "feeler gauges." The molecules used as probes must penetrate the cellulose under investigation and not be absorbed on the cellulosic surfaces. These include sugars of low molecular weight, ethylene glycols, glymes, and dextrans. Their molecular weights and diameters are given in Table 5.7. The molecular diameters of the sugars have been reported by Stone and Scallan [284].

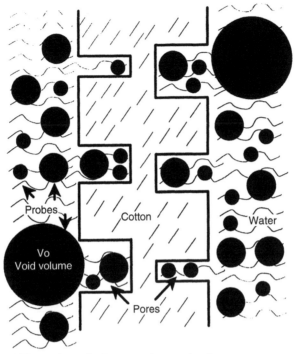

FIGURE 5.42 Simplified illustration of gel permeation mechanism.

TABLE 5.7
Molecular Probes Used in Reverse Gel Permeation Techniques for Pore Size Distribution Determination

Molecular Probe	Molecular Weight	Molecular Diameter, Å
Dextran T-40 (Void Volume)	40,000	
Sugars		
Stachyose	666.58	14
Raffinose	504.44	12
Maltose	342.30	10
Glucose	180.16	8
Ethylene Glycols		
Degree of polymerization:		
6	282.33	15.6
5	238.28	14.1
4	194.22	12.7
3	150.17	10.8
2	106.12	8.4
1	62.07	5.5
Glymes		
1	222.28	13.8
2	178.22	12.1
3	134.17	9.9
4	90.12	7.4

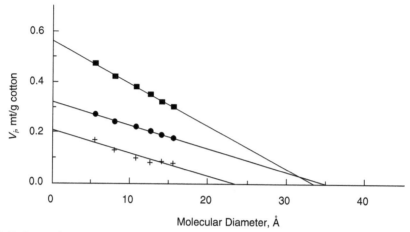

FIGURE 5.43 Internal water (V_i) that is accessible to scoured/bleached (•), caustic mercerized (■), and cross-linked with 4% DMDHEU (+) cotton cellulose. Ethylene glycols were used as molecular probes.

TABLE 5.8
Summary of Changes in Internal Pore Volume

Effect	Internal Pore Volume	Refs.
Variety	DP-90 > NX-1[a]	[289]
Temperature	Same from 30–60°C	[287]
Scouring and bleaching	Slightly increased[b]	[290]
Caustic mercerization	Substantially increased[c]	[290–292]
Liquid ammonia treatment	Moderately decreased[c] or increased[d]	[290,292,293]
Removal techniques	Moderate increases[c] from:	[293]
	Water immersion >	
	Dry, 25°C >	
	Dry, 95°C	
Cross-linking	Substantially reduced[c]	[295,297,298]
DMDHEU[e]	Reduction progressive with add-on	[295]
Formaldehyde-free agents	Less reduction than with DMDHEU	[297]
Polycarboxylic acid catalysis	More effective catalysts (e.g., NaH$_2$PO$_2$) cause greater decreases	[298]
Dyeability	DHDMI[f] cross-linking reduces internal volume but cotton remains accessible to small, but not large, direct dyes	[296]
Cellulase treatment	Mixed results	
8–14 Å pores	No change	[300]
<60 Å pores—small	Decrease in size	[301]
>60 Å pores—large	No change	[302]

[a]DP-90 is a common upland variety; NX-1 is a hybrid of Upland and Pima cottons.
[b]Relative to the original greige cotton.
[c]Relative to scoured and bleached cotton.
[d]Relative to starting purified medical cotton batting.
[e]Dimethyloldihydroxyethyleneurea.
[f]4,5-dihydroxy-1,3-dimethyl-2-imidazoildinone.

Estimates of the molecular diameters of the lower-molecular-weight ethylene glycols are based on extrapolations from measurements of Nelson and Oliver [288]. Measurements of molecular diameters were not available for the glymes but have been approximated by assuming that molecular sizes of the hydrated molecules are the same as those of the parent glycols at the same molecular weight.

The static measurement is based on the addition of a water-swollen cellulose to a solution of the molecular probe. Water in pores accessible to the solute dilutes the solution. In the chromatographic techniques, either glass or standard liquid chromatography columns were packed with cellulose in various forms. The elution volumes of the molecular probes used were determined. Data is generally plotted as internal volume accessible to individual solutes against their molecular sizes. This is illustrated in Figure 5.43.

The degree to which the internal volume has been expanded or contracted is deduced. Similar information is obtained from the static and chromatographic methods. Cellulose has been evaluated in the forms of decrystallized, ball-milled cotton, chopped cotton, cotton batting, and whole fabric.

These techniques have been used to elucidate the effects of variety [289], temperature [287], scouring–bleaching [290], caustic mercerization [290–292], liquid ammonia treatment [290,292,293], cross-linking with different agents under varying conditions [294–298], dye-ability [296,299], and treatment with cellulases [300–302], on the cotton. The trends observed are summarized in Table 5.8.

6 Chemical Properties of Cotton

CONTENTS

The cotton fiber is predominantly cellulose and its chemical reactivity is the same as that of the cellulose polymer, a β-(1→4)-linked glucan (Figure 6.1).

The chemical structure shows that the 2-OH, 3-OH, and 6-OH sites are potentially available for the same chemical reactions that occur with alcohols. If the glucan were water soluble, the

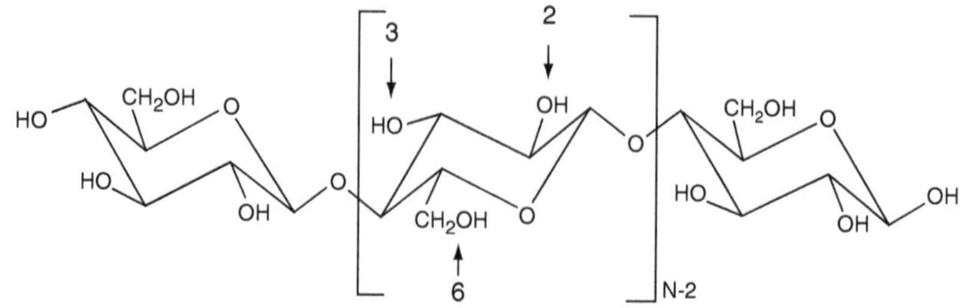

FIGURE 6.1 Ordinary reactions of chemical agents with cellulose are almost exclusively with the 2,3 and 6 hydroxyl groups that are not involved in formation of the linear polymer consisting of D-anhydroglucose units joined via β-1,4-linkages.

primary 6-OH, for steric reasons, would be the most available hydroxyl for the reaction. However, as discussed earlier, the chains of cellulose molecules associate with each other by forming intermolecular hydrogen bonds and hydrophobic bonds. These coalesce to form microfibrils that are organized into macrofibrils. The macrofibrils are organized into fibers.

The cotton fiber is subjected to many treatments that affect swelling and change its crystal structure. The agents employed must be able to interact with and disrupt the native crystalline structure in order to change it to different polymorphs. The chemical reactions of commerce generally involve the water-swollen fiber, which retains a highly crystalline structure. Reactions with this highly crystalline, water-insoluble polymer are therefore heterogeneous. Chemical agents that have access to the internal pores of the fiber find many potential reactive sites unavailable for reaction because of involvement in hydrogen bonding. Considerable light has been shed on this subject in research conducted using the CMA technique discussed earlier (Section 5.4.5). Here chemical measurements based on reaction with DEAE chloride under mild conditions showed that decreasing availability of the hydroxyl groups in cotton is 2-OH > 6-OH ≫ 3-OH. The total reactivity of the hydroxyls of cellulose and the relative reactivities of the 2-OH, 3-OH, and 6-OH differ depending on the swelling pretreatment, the reagent, and the reaction conditions. These have not been delineated for all systems.

6.1 SWELLING

6.1.1 WATER

Cellulose is hydrophilic and swells in the presence of water. Normally cellulose–water interactions are considered to occur either in intercrystalline regions or on the surfaces of the crystallites and the gross structures.

Water vapor adsorption isotherms have been obtained on cotton from room temperature up to 150°C [303,304]. Theoretical models for explaining the water vapor sorption isotherms of cellulose have been reviewed [303]. Only adsorption theories will be discussed here at ambient temperatures. The shape of the isotherm indicates that multilayer adsorption occurs and thus the Brunauer, Emmett and Teller (BET) or the Guggenheim, Anderson and deBoer (GAB) theory can be applied. In fact, the BET equation can only be applied at relative vapor pressures (RVPs) below 0.5 and after modification up to a RVP of 0.8 [305]. The GAB equation, which was not discussed in the chapter in the book *Cellulose Chemistry and Its Applications* [303], can be applied up to RVPs above 0.9 [306]. Initially as the RVP increases, a monomolecular layer of water forms in the cellulose. By a RVP of 0.19–0.22 the monomolecular layer is complete [303], and the moisture regain, when a monomolecular layer has just formed, for cotton and mercerized cotton is 3.27 and 4.56%, respectively [261,303]. By a RVP of 0.83–0.86, about three layers of water molecules are formed, and at higher RVPs it is thought that condensation occurs in the permanent capillary structure of the sample [307].

It is well known that at low moisture uptakes, the water associated with the cellulose exhibits properties that differ from those of liquid water and it has been called by such terms as "bound water," "nonsolvent water," "hydrate water," and "nonfreezing water." From a review of the literature, which included determinations by such techniques as NMR and calorimetry, Zeronian [303] concluded that between 0.10 and 0.20 g/g of the water present in the fiber cell wall appeared to be bound. Such regains are obtained at RVPs between 0.85 and 0.98.

The fiber saturation point (FSP) of cotton is the total amount of water present within the cell wall expressed as a ratio of water to solid content. It is equivalent to the water of imbibition of the fiber, also called its water retention value. The FSP has been measured using solute exclusion, centrifugation, porous plate, and hydrostatic tension techniques. It occurs at RVP greater than 0.997 and from the review of the papers, it has been concluded that the studies have yielded a value for FSP in the range of 0.43 to 0.52 g/g [303].

At equilibrium and at a particular RVP, the amount of adsorbed water held by a cellulose generally will be greater if it has been obtained following desorption from a higher RVP and not by adsorption from a lower RVP. The cause of this hysteresis is not fully established [303]. One explanation is based on the internal forces generated when dry cellulose swells, limiting the amount of moisture adsorbed whereas when swollen cellulose shrinks, stress relaxation occurs since the cellulose is plastic and permits a higher uptake of moisture.

6.1.2 SODIUM HYDROXIDE

The swelling of cotton with an aqueous solution of sodium hydroxide is an important com mercial treatment. It is called mercerization after its discoverer, John Mercer, who took a patent on the process in 1850 [308]. Other alkali metal hydroxides, notably lithium hydroxide and potassium hydroxide, will also mercerize cotton, but normally sodium hydroxide is used. Mercerization is utilized to improve such properties as dye affinity, chemical reactivity, dimensional stability, tensile strength, luster, and smoothness of the cotton fabrics [309]. The treatment is normally applied either to yarn or to the fabric itself either in the slack state to obtain, for example, stretch products, or under tension to improve such properties as strength and luster. The interaction of alkali metal hydroxides and cellulose has been exten- sively reviewed. Earlier reviews can be traced from relatively recent ones [99,310,311].

The term mercerization has to be used with care. One of the changes that occur to the treated cotton is that its crystal structure can be converted from cellulose I to II. To a researcher the term implies that the caustic treatment has induced close to complete, or full, conversion of the crystal structure to cellulose II. On the other hand, the industrial require- ment is improvement in the properties described above and these changes can be produced without full conversion of the crystal structure. For a given temperature and concentration of sodium hydroxide, the amount of swelling that occurs depends on the form of the sample. Swelling deceases in the order fiber > yarn > fabric. In addition, properties are affected by whether the material is treated in slack condition or under tension. Finally, depending on the processing time, the material, and other conditions (e.g., caustic concentration, temperature, slack or tension treatment) mercerization, as defined by researchers, might not extend beyond peripheral regions. In mercerizing fabrics industrially, the following variables need to be considered: caustic strength, temperature, time of contact, squeeze, framing and washing, and the use of penetrants. Abrahams [309] has provided the following guidelines. The caustic concentration should preferably be in the range of 48–54°Tw (approximately 6.8 to 7.6 M), although if improved dye affinity is the objective 30–35°Tw (roughly 4.0 to 4.7 M) can be used. Temperature may vary in the range of 70–100°F (21.1 to 37.8°C) at higher concentrations but has to be monitored more closely when the concentration is 30°Tw. A contact time of 30 sec can be used normally. A penetrant is essential if the fabric is in the greige state to permit wetting. Washing on the frame is enhanced by using a mercerizing penetrant that is an active detergent over a wide caustic range.

Hot mercerization allows better penetration of the alkali into the fibers than the ambient temperatures used normally [312]. However, to obtain optimum improvement in properties the caustic has to be washed out after the fabric is cooled.

During mercerization, the swelling induced by the caustic is inhibited from outward expansion by the presence of the primary wall of the cotton fiber. The changes observed in fiber morphology by mercerization include deconvolution, decrease in the size of the lumen, and a more circular cross section.

Changes in the fine structure that occur when cotton is mercerized include a conversion of the crystal lattice from cellulose I to II, a marked reduction in crystallite length, a marked increase in moisture regain, and a reduction in degree of crystallinity [99,311]. A higher concentration is required to induce the optimum changes as the temperature is increased from subambient to room temperature. The conversion from cellulose I to II is substantially complete in cotton yarn treated at 0°C for 1 h, with 5 M LiOH, NaOH, or KOH [313]. In the case of the sample treated with 5 M NaOH, the following changes were noted: the extent of swelling, measured by the 2-propanol technique, roughly tripled; the moisture regain increased by about 50%; and the crystallite length decrease by approximately 40% [313]. An estimate of the loss in crystallinity on mercerization, determined by moisture regain measurements, can be found in Table 5.2.

The effect of mercerization on tensile properties depends on the type of cotton tested. In one study six *G. barbadense* samples were slack mercerized and the breaking forces and tenacities of the fibers relative to their nonmercerized counterparts ranged from 88 to 122% and from 80 to 114%, respectively [314]. A larger change was found in the case of a *G. hirsutum* Deltapine Smoothleaf sample. In this case, the relative breaking force and tenacity were 186 and 134%, respectively. Relative breaking strains ranged from 160 to 189% for the *G. barbadense* samples and 150% for the *G. hirsutum*. The increased strength and extensibility of slack mercerized cotton have been attributed partly to the deconvolution that has occurred and partly to the relief of internal stresses [315]. The reduction in crystallinity and crystallite length that results from mercerization contributes to the relief of stresses in the fiber as well as in giving a product of higher extensibility.

There is some evidence that the degree of hydration of alkali hydroxide ions affects their ability to enter and swell cellulose fibers [310]. At low concentrations of sodium hydroxide, the diameters of the hydrated ions are too large for easy penetration into the fibers. As the concentration increases, the number of water molecules available for the formation of hydrates decreases and therefore their size decreases. Small hydrates can diffuse into the high order, or crystalline regions, as well as into the pores and low-order regions. The hydrates can form hydrogen bonds with the cellulose molecules.

Ternary complexes called soda celluloses can form between cellulose, sodium hydroxide, and water [310]. In these complexes, some of the water molecules of the sodium hydroxide hydrates are replaced by the hydroxyl groups of the cellulose [310]. The x-ray diffraction diagrams have been obtained for five soda celluloses as intermediates in the formation of cellulose II from cellulose I [310,316].

6.1.3 LIQUID AMMONIA

Another swelling reagent for cotton cellulose, which is also used industrially, is liquid ammonia. This treatment has been extensively reviewed and discussed [311,312,317–321]. Anhydrous ammonia penetrates the cellulose relatively easily and reacts with the hydroxyl groups after breaking the hydrogen bonds. The reaction occurs first in the LORs, and gradually later in the crystalline regions of the fibers. An intermediate ammonia–cellulose (A–C) complex, held together by strong hydrogen bonds, is formed. This complex can decompose in several ways, and yield different products, depending on the condition of the removal of the ammonia. Lewin and Roldan [319] developed a phase diagram (Figure 6.2) of the four major phases represented in the four corners of a tetrahedron [319].

The directions of the transitions between the various phases are indicated by the arrows, i.e., a transition from D to III is possible on application of dry heat. A transition from III to D is impossible unless a strong swelling agent like ammonia is used. A transition from III to I is possible by the application of water and heat or by a prolonged application of water at ambient conditions. The reverse transition is impossible without an intermediate swelling step. The transitions are usually not complete, especially in industries, and a wide range of products can be obtained as indicated by the phase diagram. The ammonia–cellulose complex and cellulose III can also be obtained from cellulose II. There is, however no reversion to cellulose I.

The CI decreases upon liquid ammonia treatment and rinsing with water with or without heating, from 79 to 30–40 Å, and the crystallite size decreases from 54 to 37–34 Å [319]. The circularity and homogeneity are also increased. The tensile strength is greatly increased and the elongation is decreased upon stretching the ammonia-treated fibers. The accessibility and consequently the dyeability of the fibers are also greatly increased.

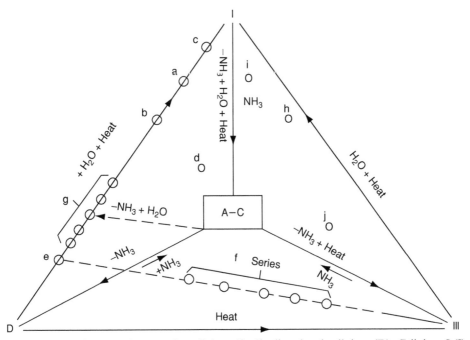

FIGURE 6.2 Phase diagram of ammonia–cellulose (A–C), disordered cellulose (D), Cellulose I (I) and Cellulose III (III). A–C is the vertex of a tetrahedron and is placed above the plane of the paper. The various samples studied are placed in the basal plane. Arrows show the transition directions.

The interactions between cellulose and ammonia have attracted industrial attention. Applications to wood [321–325] by treating it with liquid ammonia and with gaseous ammonia under pressure [326] to cotton fabrics [320,325–327], and to sewing treads [327] have been described.

It has been suggested that the great depth of color or dye yield found with mercerized cotton is due to the caustic treatment inducing an abundance of large pores in the fiber. In contrast, the high level of resilience associated with liquid ammonia treatment has been ascribed to a low level of large pores in the fiber [292].

6.2 ETHERIFICATION

6.2.1 GENERAL

Cellulose ethers generally are very stable. Many etherified cottons are highly resistant to hydrolytic removal of substituent groups under both acidic and alkaline conditions. Because of this stability, many of the most practical chemical treatments of cotton are based on etherification reactions [9,328–331]. These treatments provide cotton products with useful, durable properties including wrinkle resistance, water repellency, flame resistance, and anti-microbial action.

Treatments based on condensation reactions (such as in the classical Williamson synthesis) produce the most stable cotton derivatives. On the other hand, treatments based on addition reactions (such as in the Michael reaction) yield cellulose ethers that are somewhat less stable. This lower level of stability is because of the equilibrium nature of the addition reaction. Typical examples of these two types of cellulose etherification are carboxymethylation [9,329,330] and cyanoethylation [9,329,330,332], respectively, both of which proceed in the presence of alkali.

In carboxymethylation of cotton, the fibers are impregnated with aqueous sodium hydroxide and then treated with chloroacetic acid:

$$Cell + 2NaOH + ClCH_2COOH \rightarrow Cell-OCH_2COONa + NaCl + 2H_2O \qquad (R6.1^*)$$

The equation for the cyanoethylation of cotton with acrylonitrile can be represented as:

$$Cell-OH + CH_2{=}CHCN \rightleftharpoons Cell-OCH_2-CH_2-CN \qquad (R6.2)$$

Cotton etherification can be carried out so that the cotton fiber retains its fiber structure and textile properties. The latter, however, often are greatly modified, even at low degrees of substitution (DS). It is possible, through etherification, depending upon the ether group substituted and the DS on the cellulose, to obtain fibrous products with widely diverse properties. For example, there are cellulose ethers that have very high levels of moisture absorbency or even water solubility (carboxymethylated cotton) and others that have high levels of water repellency (long-chain alkyl ethers and stearamidomethyl ethers).

A generalized representation of cellulose ethers is Cell—OR, wherein the ether group (R) is alkyl, aromatic, heteroalkyl, heterocyclic, or other substituent, including ether groups bearing other functional groups. Cellulose ethers with mixed ether substituents also have been prepared by treatment with two or more reactants, either in combination or in sequence.

---

*R denotes reaction.

The extent to which the cellulose molecule is modified is denoted as the DS. Thus, the DS is the average number of alcohol (OH) groups of the anhydroglucose unit of the cellulose molecule that have been substituted. The DS can range to a value of 3.0, which indicates that all the OH groups of the anhydroglucose units that make up the polymeric cellulose molecule have been etherified. A DS of 0.25 indicates that, on an average, there is 0.25 ether group per anhydroglucose unit, or alternatively stated, there is an average of one ether group for every four anhydroglucose units.

It must be noted that in cotton not all of the OH groups are accessible for reaction and substitution because of the unique polymeric structure of the cotton fiber. Thus, the average DS of the etherified cotton usually signifies a higher DS in the accessible regions and no substitution in the inaccessible crystalline regions of the fiber. It must be remembered that because of the heterogeneous nature of cotton we are always speaking of the average DS of the cotton product.

The reaction between cotton and ethylene oxide to give hydroxyethylated cotton appears to be simple and straightforward:

$$\text{Cell}-\text{OH} + \underset{\underset{\displaystyle O}{\diagdown\diagup}}{\text{CH}_2\text{CH}_2} \rightarrow \text{Cell}-\text{OCH}_2\text{CH}_2\text{OH} \tag{R6.3}$$

However, once introduced on the cellulose molecule, the hydroxyethyl group is highly reactive and capable of further reaction with another molecule of ethylene oxide. This type of reaction, called graft etherification, can build up relatively long-chain substituents on the cellulose molecule without involvement of many additional cellulosic OH groups:

$$\text{Cell}-\text{O}(\text{CH}_2\text{CH}_2\text{O})_n\text{H} \tag{R6.4}$$

In these products, simple analysis gives the average molar substitution (MS) per anhydroglucose unit of the cellulose rather than the DS. Specialized analytical techniques must be employed to ascertain the DS of the grafted derivatives.

6.2.2   WRINKLE RESISTANCE

The most important cotton etherification treatments are those that produce wrinkle resistance in fabrics [331,333,334]. The aldehydes, formaldehydes, and glyoxals, react with the OH groups of two cellulose chains as well as those of one chain. Reaction in which a bond is established between the two cellulose molecules is called cross-linking and is the basis for profound changes in the cotton fiber. Cross-linking produces resiliency in the fiber to give the needed dimensional stabilization, wrinkle resistance, and crease retention for modern durable-press cellulosic fabrics. Cross-links based on etherification reactions traditionally have been used because of their durability to repeated laundering and wear.

Methylenation of cotton by treatment with formaldehyde has been an elusive objective of textile finishers for almost 100 years:

$$2\text{Cell}-\text{OH} + \text{HCHO} \leftrightarrow \text{Cell}-\text{OCH}_2\text{O}-\text{Cell} + \text{H}_2\text{O} \tag{R6.5}$$

Although formaldehyde is inexpensive, readily available, highly reactive, and ideally would be the simplest ether cross-link between cellulose chains, there has been only limited successful usage of this reagent to produce wrinkle resistant cotton. A treatment based on gaseous- or vapor-phase application of formaldehyde to cotton under rigidly controlled conditions has gained some acceptance.

Methylolamide agents are most commonly used to produce wrinkle resistance and dimensional stabilization in cotton [331,333–337]. These agents, formaldehyde adducts of amides or amide-like nitrogenous compounds, introduce ether cross-links between cellulose molecules of the cotton fiber. Chemical processing of cotton with methylolamide agents is the most widely practiced textile finishing treatment throughout the world.

The first methylolamide agent for cotton was the urea–formaldehyde adduct. Today, most finishing of cotton with methylolamide agents uses cyclic urea–formaldehyde adducts. The most commonly used agent is dimethyloldihydroxyethyleneurea (DMDHEU):

$$\begin{array}{c}
\text{O} \\
\|\\
\text{HO}-\text{H}_2\text{C}-\text{N} \qquad \text{N}-\text{CH}_2-\text{OH} \\
\\
\text{HO} \qquad\qquad \text{OH}
\end{array}$$

The chemical name of this reagent is 1,3-bis(hydroxymethyl)-4,5-dihydroxy-imidazolidinone-2 but it is usually called DMDHEU or the glyoxal reactant because it is prepared from glyoxal, urea, and formaldehyde. Other methylolamide agents that have been used for producing wrinkle resistance in cotton include the aforementioned urea–formaldehyde, dimethylolurea, dimethylolethyleneurea, and formaldehyde adducts of melamines (triazines), acetylenediurea, propyleneurea, uron, triazones, and alkyl carbamates. Reactions between methylolamides and cellulose occur in the presence of acid (or Lewis acid) catalysts and are very fast at elevated temperatures—sufficiently so that they are adaptable to the requirements of rapid, commercial processing of cotton fabrics.

6.2.3 Reactive Dyeing

Another important commercial utilization of cotton etherification is in coloration of fabrics with reactive dyes [338–340]. Reactive dyes contain chromophoric groups attached to moieties that have functions capable of reaction with cotton cellulose by nucleophilic addition or nucleophilic substitution to form covalent bonds. In the nucleophilic addition reaction, an alkaline media transforms the reactive dye to an active species by converting the sulfatoethylsulfone to the vinyl sulfone, which reacts with cellulose to form an ether bond. In the nuceophilic substitution reaction, a halogen atom on the reactive dye molecule is replaced by an oxygen cellulose ester covalent bond. Reactive dyes that have been commercialized include monofunctional dyes with the following reactive groups: mono- and dichlorotriazine, sulfatoethylsulfone, sulfatoethylsulfonamide, trichlorpyrimidine, dichloroquinoxaline, difluoropyrimidine, and difluorochloropyrimidine; and bifunctional dyes (contain two different reactive moieties on the dye molecule) with the following reactive moieties: bis aminochlorotriazine, bis aminonicotinotrazine, aminochlorotriazine–sulfatoethylsulfone, and aminoflorotriazine–sulfatoethylsulfone [341].

Reactive dyes with methylolamide-like groups were used on cotton at one time [342]. Bonding to cellulose was similar to that in etherification treatments to produce wrinkle resistance. However, because of technical problems in their application, usage of these formaldehyde-based reactive dyes has essentially ceased. Fixatives are used, which act through methylol groups, to improve color fastness of direct and other dyes on cotton. Their mechanism includes bonding (etherification) between dye and cellulose as well as between dye molecules.

6.2.4 FLAME RESISTANCE

Improved flame resistance is an important and useful property that can be imparted to cotton fibers and cotton textiles. The flammability and flame retardancy of cotton have been studied extensively and several comprehensive reviews are available [343–346].

Once ignited, virtually all common textile fabrics will burn. Textile fabrics burn by two distinctly different processes: flaming combustion and smoldering combustion. As the fibers that make up fabrics are composed of large, nonvolatile polymers, flaming combustion (e.g., that caused by an open-flame source, such as a match) requires that the polymer undergo decomposition to form the small, volatile organic compounds that constitute the fuel for the flame. The combustion of polymers is a very complex, rapidly changing system that is not yet fully understood. For many common polymers, this decomposition is primarily pyrolytic with little or no thermo-oxidative character. Smoldering combustion (e.g., that caused by a cigarette) on the other hand involves direct oxidation of the polymer and chars and other nonvolatile decomposition products. Since smolder ignition and open-flame ignition are different mechanisms, they usually require different flame retardant (FR) treatments and treatments to control open-flame ignition can adversely affect smolder resistance of 100% cellulosic and predominately cellulosic fabrics [347].

The flame resistance of a textile is test method dependent, i.e., it should be specified what test the material passes when making a claim of flame resistance. There are currently many mandatory U.S. federal and state and international standards, as well as voluntary national and international standards (smolder or cigarette resistance; small and large open flame), for the flammability of textiles. There are component standards and composite or full scale standards. The current smolder test methods use a standard cigarette. Large open-flame standards for mattress and box springs use burners that mimic burning bedclothes; other large flame sources use, e.g., an 18 kW flame for 3 min, or a trash can full of burning paper. Small open-flame sources usually are matches, cigarette lighters, and candles (e.g., a butane or a propane flame for 15 or 20 s). For general wearing apparel, there is a 45° angle test with a small open flame and burn rate is measured. For children's sleepwear, the test is a vertical flame test with a small open flame and the distance the flame travels is measured.

Whenever cotton is ignited in the presence of oxygen and the temperature is high enough to initiate combustion, untreated cotton will either burn (flaming combustion) or smolder (smolder combustion means burning and smoking or wasting away by a slow and suppressed oxidation without flame) until carbonaceous material or combustion gases result (see Section 6.4.6). Ignition (defined as initiation of self-sustaining flaming combustion for an observable time) occurs as a result of exothermic reactions between volatile decomposition products and oxygen [348]. The major factors that influence ignition of cellulosic materials like cotton are air flow, relative humidity of the fabric, the amount of oxygen available, physical factors (geometry, density, thickness, etc.) [349,350], chemical factors (e.g., inorganic impurities) [351–354]), heat source, and how fast the cotton is heated [350]. Because of these variables, it is not really correct to talk about a specific ignition temperature or autoignition temperature (defined as the temperature under controlled conditions in air when there is a runaway weight loss and flaming combustion resulting from contact with heated air in the absence of any spark or flame) for cotton. Thermal analysis studies in air and in 8.4% oxygen indicate that cotton and pure cellulose ignite at about 360 to 425°C [348,355,356]. Cellulose does not melt but is reported to decompose (char) at 260–270°C [357] and cotton at 338°C [348].

Numerous end uses for cotton depend on its ability to be treated with chemical agents that confer flame resistance. The chemical FR treatments used to confer flame resistance to untreated cotton depend on many factors [358]: is the finish intended to be durable or nondurable; is the treatment used is to prevent burning or smoldering; what is the construction of the textile to be treated; and is the textile 100% cotton or does it contain some percentage of thermoplastic human-made fibers, i.e., polyester? In addition, problems of toxicological and ecological concerns have more recently assumed significance [359]. Thus, it is clear that the process of treating and finishing cotton fabrics to make them flame resistant is complex and may be relatively expensive.

6.2.4.1 Nondurable Finishes

A number of nondurable finishes have been developed, which are in use in industrial cotton fabrics. Most are based on borax ($Na_2B_4O_7 \cdot 10H_2O$), boric acid (H_3BO_3), diammonium phosphate (($NH_4)_2HPO_4$), or sodium phosphate, dodecahydrate ($Na_3PO_4 \cdot 12H_2O$) [360]. These agents are applied to the textile from water solutions, followed by squeezing to reduce the wet pick up and finally by drying. It should be realized that although such treatments are removed by conventional laundering, most of them could withstand several nonaqueous launderings with dry cleaning solvents and remain effective. These nondurable finishes are recommended for 100% cotton textiles. For cotton–polyester blend textiles, a reagent such as ammonium bromide, which decomposes on heating and becomes active in the gaseous phase, should be added to the agents referred to above for greater effectiveness.

6.2.4.2 Durable Finishes

Many durable flame retardants for cotton have been developed to convey open-flame resistance [344,346,360,361]. The vertical flame test for determining the U.S. children's sleep-wear flammability (16 CFR 1615 and 1616) is a rather severe test and cotton fabrics require a FR treatment to pass the test. The test method requires treatments that are durable to 50 hot water wash and dry cycles. Currently there are relatively few commercially available FR chemistries that are durable under these conditions required today. Some of the reasons include low commercial availability of the chemicals, costs, safety concerns, process control issues, and difficulty in application.

Treatments with more limited durability would be more appropriate for the U.S. general wearing apparel (16 CFR 1610), which has a 45° angle test because the test and the cleaning requirements are less severe (see Section 6.2.4.4). Currently only resistance to dry cleaning and hand washing are required.

The main durable FR finishes used on cotton to meet more severe open-flame resistance requirements are phosphorus based [343,358]. One of the problems with typical phosphorus-based FR treatments on fleece, which only requires a mild treatment to pass the 45° angle test, is that the often-required levels alter the esthetic properties of the fleece, resulting in a fabric that is stiff or matted and often has unpleasant odors. Most common types of dyes used on cotton are affected by pH or oxidation–reduction procedures that are used during the FR treatments.

The most successful durable FR finish for open-flame resistance is based on the neutralized moiety of tetrakis(hydroxymethyl)phosphonium chloride (THPC), i.e., (THPOH). When the THPOH is applied to cotton fabric as described previously, the dried fabric can be cured with ammonia gas to form the water-insoluble THPOH–NH_3 polymer [362]. The resin polymerizes within the cotton fibers to form an insoluble polymer that withstands many repeated aqueous launderings. A THPOH precondensate applied to cotton fibers along with ammonia (precon-densate–NH_3 process) is currently the commercial system used successfully to impart durable flame resistance to cotton garments (e.g., safety apparel). Durability and freedom from fabric odor is greatly improved by peroxide oxidation of the ammoniated fabric [360]. The tendency of THPOH to release formaldehyde during drying can be a potential health problem for textile workers during finish applications (see Chapter 10). This has been largely solved by the reaction of urea with the phosphonium salt prior to application to the cotton textile and by more efficient hooding of the drying section of the process [363].

Other phosphorus-containing flame retardants achieve their durability by reacting with the cellulose molecules and polymerizing within cotton fibers. Reactive phosphorus-based flame retardants (e.g., phosphonic acid ester) are typically applied by a pad–dry–cure method in the presence of a phosphoric acid catalyst. Examples of such flame retardants are N-methylolamides of phosphines and phosphine oxides, and vinyl phosphonates. Nitrogen-containing agents used with these phosphorus-containing agents include urea, N-methylol-propionamide, and methylolated melamines.

Although the precondensate–NH$_3$ process has been the most successful FR treatment for 100% cotton textiles, it is not completely effective when the cotton textile contains appreciable amounts of thermoplastic synthetic fibers like polyester [364,365]. Extensive research has been carried out on the treatment of cotton–polyester blends with flame-retardant finishes and it has been found that a dual treatment (i.e., one for the cotton component and one for the polyester) is the most successful [360]. The best finish for the cotton component is precondensate–NH$_3$. The best agent for rendering the polyester component flame retardant was tris(2,3-dibromopropyl)phosphate (Tris) [366]. This chemical was found to be a mutagen–carcinogen, and in 1977, the Consumer Product Safety Commission banned its use on children's sleepwear. The children's sleepwear industry responded by marketing 100% thermoplastic fabrics (i.e., polyester and nylon) without any chemical treatment. These materials melt at relatively low temperatures before the fabric ignition temperature is reached. There is still a problem of skin injury from the contact of the molten synthetic component with the skin, but this was deemed less serious than the possible exposure to a known carcinogen, i.e., Tris. The U.S. children's sleepwear flammability standard test method (16 CFR 1615 and 1616) was amended in 1978 so that fabrics no longer had to be treated to control for the melt drip phenomenon. These standards were further amended in 1996 to not require the vertical flame test for tight-fitting garments and for infant wear (0–9 months). Currently the most successful FR finish for cotton–polyester fabric involves the application of antimony–halogen finishes [367]. The finish is expensive and difficult to apply, but the results are good, even when the fabric is intended for outdoor use.

Research has also been done on the FR treatment of cotton blend fabrics containing a flame-resistant synthetic fiber, such as a FR-modacrylic, or an aramid. A FR finish must still be applied to the cotton component. Thus so far, research has proved that these treated blends are more successful with heavier-weight goods, such as those intended for military applications. The observation that heavier-weight goods are easier to treat for flame resistance than their lightweight counterparts holds true for almost all FR finishes on almost all cotton containing fabrics. The best finish for durable fire retardant properties on lightweight cotton goods (i.e., 4 oz/yd^2 or less) is precondensate–NH$_3$ process.

Backcoating of upholstery fabrics, using decabromodiphenyl ether (deca-BDE) or hexabromocyclododecane, antimony trioxide, and an acrylic latex, is effective for flame resistance, but there are toxicity concerns. The use of synthetic barrier fabrics (i.e., interior fire-blocking) may remove the need for chemical backcoating.

Another approach to flame-resistant cotton containing fabrics involves the use of core spun yarns [368–372]. There are two components in these specialized yarns. One component is a central core usually made from a human-made polyester or nylon, or a nonflammable core like fiberglass. The other component is a cotton cover that is wound around the central core to form the core yarn. The core yarn is woven or knitted into an appropriate textile, then treated with a finish to make the flame-resistant cotton cover. When the core yarns are spun to restrict their synthetic content to 40% or less, the FR treatment of the cotton component alone will frequently make the array flame resistant. The need for a separate FR treatment of the polyester or nylon component is no longer required.

6.2.4.3   Cotton Batting or Filling

For upholstered furniture, mattresses and foundations, and filled bedclothes, there is a need for both smolder-resistant and open flame-resistant cotton batting. Alkali metal ions on fabrics and batting (natural in raw cotton and from finishing agents) induced smoldering of cotton [351,352]. Postrinsing of upholstered fabrics or washing cotton batting does not completely remove the alkali metal ions and could cause a multitude of technical problems [373]. Thermo-plastic fiber in blended cotton fabrics and batting appears to convey cigarette or smolder resistance until the cotton content exceeds about 80%. Most treatments that control for flame resistance do not control for smolder resistance [347]. Nondurable agents such as the boric acid, borax, and ammonium phosphate have been used to inhibit the smolder of cotton and to pass open-flame mattress and upholstered furniture flammability requirements. Research into converting these treatments into more durable finishes also has been investigated. Incorporation of trimethylolmelamine (TMM) into a borax-type finish seemed promising [374], but ecological problems with TMM slowed its widespread application.

Engineered cotton batting properly treated with boric acid (~10% or greater) and blended with inherently flame-resistant fibers (e.g., enhanced FR-modacrylic, FR-PET, Visil (silica containing cellulosic), etc.) is cigarette (smolder) resistant and open flame-resistant and can be used like any other padding material [375]. It is an improved product that can be used as a drop-in component fire-blocking barrier in the mainstream soft furnishings (i.e., mattresses, beddings, and upholstered furnitures). Ammonium phosphate-treated cotton, blended with FR-polyester or FR-modacrylic, is also used to make flame-resistant and smolder-resistant engineered batting to meet the flammability requirements of the mattresses and upholstered furnitures. Cotton batting—boric acid or ammonium phosphate treated or blended—should be helpful in meeting the various cigarette resistance and open-flame resistance regulations for mattresses, futons, and upholstered furniture.

6.2.4.4   Polycarboxylic Acids as Flame Retardants

Carpet materials larger than 1.2×1.8 m must pass a standard test method (referred to as the pill test; 16 CFR 1632) to be used commercially. Some untreated high-density cotton-containing materials pass this test, but the risk of failure is too high without further modification to improve the flame resistance. To be of commercial use chemical modification of cotton-containing carpets must be devoid of formaldehyde or other harmful substances, must not discolor the surface, must not cause dye shade change, must undergo reaction at less than 150°C to avoid synthetic backing deformation, must be applied at low moisture contents, and must be cost effective. An effective solution has come from an unexpected place. Polycarboxylic acids, such as 1,2,3,4-butanetetracarboxylic acid (BTCA), and citric acid catalyzed by sodium hypophosphite or sodium phosphate that have been shown to be effective durable-press agents for cotton fabrics [376], are also effective for improving the flame resistance of carpets and other cotton textiles that require a mild FR treatment to pass the appropriate flammability test for that product. These are not considered FR agents but these polycarboxylic acid systems were shown to be capable of providing cotton-containing carpeting with flame suppression properties sufficient to pass the methenamine pill test [377–379]. This is attributed to increased char formation upon combustion. A similar approach with polycarboxylic acids has been used for raised-surface apparel [380].

6.2.5 MULTIFUNCTIONAL PROPERTIES

Etherification of cellulosic fibers by N-methylol groups in cross-linking resins usually occurs directly by reaction with the various hydroxyl groups in the anhydroglucose unit. However, when polyethyleneglycols are present, the semicrystalline polyols are preferentially etherified by the N-methylol groups at very low curing temperatures (as low as 80°C) and some grafting occurs on the cellulosic hydroxyl groups. The net effect is a flexible composite structure of the cross-linked polyol as an elastomeric coating on cotton fibers or fabrics.

The resultant fabrics are unique in that they have many functional property improvements: thermal adaptability due to the phase change nature of the bound polyol, durable press or resiliency, soil release, reduction of static charge, antimicrobial activity, enhanced hydrophilicity and improved flex life, and resistance to pilling. Because of the different molecular weights of polyols, resins, acid catalysts, and fabric constructions, there are numerous modified fabrics that can be produced with sets of improved attributes. Each fabric must be carefully evaluated for optimum curing conditions and formulations to produce the desired product. Several licenses have been granted for this process. Various types of apparel, healthcare items, and industrial fabrics are currently evaluated for commercial production [381,382].

6.3 ESTERIFICATION

6.3.1 GENERAL

Hydroxyl groups of cotton cellulose can be reacted with carboxylic acids, acyl halides, anhydrides, isocyanates, and ketenes to produce cellulose esters retaining the original fiber, yarn, and fabric form of the textile material. Prior to the reaction, native cotton exhibits a high degree of crystallinity, unusually strong and extensive hydrogen bonds, and a cellulose molecular weight exceeding 1 million. These properties greatly limit accessibility of the cotton fiber's interior to penetration by chemical agents. Tightness of yarn and fabric construction also hinders esterification to high DS. Pretreatment with swelling agents is usually required for esterification with monofunctional carboxylic acids and anhydrides. Strong acid catalysts may be used just as in esterifying simple alcohols. However, acid-catalyzed esterification of cotton must be at or below room temperature to prevent chain cleavage of the cellulose, and as a result, 1–3 h may be needed to reach the desired DS.

6.3.2 ACETYLATION

Acetylation of spun or woven cotton to an acetyl content of 21% results in a DS of 1.0 acetyl group per anhydroglucose unit (AGU), and imparts rot resistance as well as heat or scorch resistance to yarn or fabric. The partially acetylated cotton has been used commercially as a covering material for ironing boards and hothead laundry presses. Fully acetylated cotton has a DS of 2.7–3.0 acetyl groups per AGU. It has even greater rot and heat resistance, and is sufficiently thermoplastic to permit permanent creasing and pleating at conventional ironing temperatures. It is highly resistant to organic solvents, in contrast to low-molecular-weight cellulose triacetate, but readily accepts the disperse dyes used with acetate rayon.

6.3.3 FORMALDEHYDE-FREE WRINKLE RESISTANCE

The high-temperature cross-linking of cotton cellulose by polycarboxylic acids, having three to four carboxyls per molecule, has been extensively investigated as a method of formaldehyde-free durable-press finishing. In 1963, Gagliardi and Shippee [383] showed that polycarboxylic acids are capable of imparting wrinkle and shrinkage resistance to cellulosic fabrics such as cotton, viscose rayon, and linen. Citric acid was the most effective agent tested but it did cause more fabric discoloration than other polycarboxylic acids. In 1963, progress was made with the use of alkaline catalysts such as sodium carbonate or triethanolamine [384]. Acids having four to six carboxylic groups per molecule were usually more effective than those having two or three carboxyls. These finishes were recurable as the ester cross-links appeared to be mobile at high temperature. Acids investigated were *cis*-1,2,3,4-cyclopentanetetracarboxylic acid and BTCA.

Later a breakthrough came when a series of weak base catalysts were discovered that are more active than sodium carbonate or tertiary amines. Alkali metals salts of phosphoric, polyphosphoric, phosphorous, and hypophosphorous acids were proven effective [376,385–389].

Subsequent to this breakthrough, the subject was extensively investigated and was subsequently reviewed [390]. The acids most effective are BTCA, tricarballylic acid, and citric acid. Owing to the low cost and wide availability of citric acid, it is undergoing widespread commercial development for esterification cross-linking of cotton and paper products, often with minor amounts of BTCA as an activator, together with suitable catalysts.

Pad-dry-heat curing technology suitable for use in continuous mill processing of cotton fabric has been demonstrated with BTCA and citric acid, using weak bases as high-speed curing catalysts. In order of increasing effectiveness as catalysts are the sodium salts of phosphoric, polyphosphoric, phosphorous, and hypophosphorous acid, most of which are superior to sodium carbonate or tertiary amines in catalytic activity. Heat curing is carried out at 160–215°C for 10–120 s. Smooth drying properties so imparted are durable to repeated laundering with alkaline detergents at 60°C (140°F) and pH 10. The surprising degree of resistance of the ester groups to alkaline hydrolysis is apparently due to the presence of unesterified carboxyl groups in the fabric finish, which are converted to negatively charged carboxylate ions by reaction with alkaline detergent. These negatively charged fibers tend to repel hydrolytic attack on ester groups by carbonate, phosphate, and hydroxide ions of the detergent. The carboxylate ions of the finish strongly adsorb cationic dyes, and impart soil release properties.

The mechanism of high-temperature, base-catalyzed esterification involves cyclic anhydride formation by the polycarboxylic acid, followed by base-catalyzed reaction of anhydride groups with cellulosic hydroxyls. Fumaric acid, which cannot form a cyclic anhydride because of the *trans* arrangement of the two carboxyl groups, fails to react with cotton in the presence of weak bases and heat.

As formaldehyde-free, durable-press finishing agents, the polycarboxylic acids are apparently free of adverse physiological effects noted for diepoxides, vinyl sulfone derivatives, and polyhalides previously proposed for this purpose. In addition to imparting shrinkage resistance, smooth drying, and durable-press properties, ester-type cross-linking has undergone commercial development for treatment of woven and nonwoven cellulosic materials used in diapers, sanitary napkins, and other multilayer personal care products where resiliency and recovery from wet compression are required.

6.3.4 Esters of Inorganic Acids

Esters of cotton cellulose with inorganic acids have also been prepared in fiber, yarn, or fabric form. Nitric acid can be reacted with cotton linters to produce explosive cellulose nitrates known as guncotton, which contains more than 13% nitrogen, and is used in smokeless powder. The FR properties can be imparted to cotton fabric by oven curing with mono- or diammonium phosphate at high temperature in the presence of urea as a catalyst and coreactant. The resulting ammonium cellulose phosphate is flame- and glow-resistant, but ion exchange with sodium salts or especially calcium salts, during fabric laundering suppresses these properties.

Aftertreatment of phosphorylated fabric in its ammonium or sodium ion form with aqueous titanium sulfate produces a finish having increased resistance to ion exchange with calcium or magnesium salts. The flame and glow resistance obtained are fairly durable to repeated laundering. Sulfation of cotton with ammonium sulfamate (AS) in the presence of urea or urea-based cross-linking agents imparts an excellent flame resistance to cotton that is durable to over 50 alkaline launderings in soft as well as in hard water [391–393]. When bis(hydroxy-methyl)uron was used as cross-linking agent together with AS, the durable-press rating increased from 1.0 for the untreated fabric to 2.5–3.0 after 25 launderings. The severe afterglow of the sulfated fabrics could be overcome either by an aftertreatment with diammonium phosphate, which was not durable, or by a combined and simultaneous sulfation and phosphorylation using AS and phosphorus triamide (PA) or methyl substituted PA in the ratio of P:S of 1.3:1.0. The tensile and tear strengths of the phosphorylated and sulfated fabrics are decreased by about 30–40%, whereas the stiffness and air permeability remain nearly the same. The treated fabrics stood up to 25 hard water launderings with chlorine bleaches. Sulfamide, $SO_2(NH_2)_2$, cross-links cotton and imparts flame resistance when cured on fabric at high temperatures in the presence of urea. The flame resistance is durable to laundering with ionic detergents, but resistance to afterglow is lacking unless nonionic phosphorus-containing flame retardants are also applied [376,388,389,391,394–401].

6.3.4.1 Bioconjugation of Esterified Cotton Cellulose

Biologically active conjugates of cotton cellulose have been synthesized using glycine ester-ification as a starting point in the synthesis (Figure 6.3).

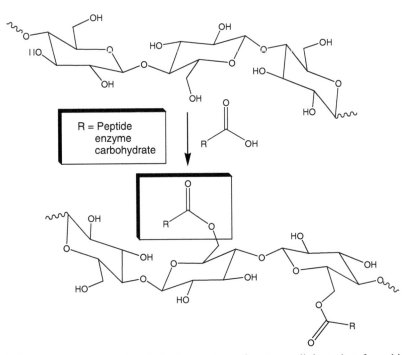

FIGURE 6.3 Peptide, enzyme, and carbohydrate esters of cotton cellulose that form bioconjugates creating biologically active cotton textile surfaces.

Peptides and proteins were synthesized or immobilized on cotton to create biologically active fibers with protease sequestering and antibacterial activity. Glycine esterified to cotton cellulose was employed as a linker to assemble peptide sequences and immobilize enzymes on fabrics [402]. Esterification of cotton cellulose with glycine was accomplished using a dimethylamino-pyridine-catalyzed, carbodiimide–hydroxybenzotriazole acylation reaction. When the base-catalyzed esterification of cotton cellulose was compared between cotton twill fabrics and cellulase-treated cotton twill fabrics dimethylaminopyridine-cataylzed esterification of cotton gave higher levels of glycine esterification in the cellulase-treated samples. Peptide substrates of human neutrophil elastase (HNE) were synthesized on glycine esterified cotton cellulose to demonstrate the use of HNE substrates as protease sequestering agents of HNE covalently attached to cotton fibers. The HNE substrate peptide conjugate on cotton cellulose has served as a route to demonstrating the efficacy of removing proteases from the chronic wound. This approach is also a model for the design of peptidocellulose analogs in wound dressing fibers for chronic wounds. It has lead to the development of interactive cotton-based wound dressings that redress the proteolytic imbalance, resulting from the inflammatory pathophysiology of the chronic wound [403,404]. In analogous approaches, glycine esterified cotton cellulose was employed to immobilize lysozyme on cotton twill fabric [405]. Lysozyme was immobilized on glycine-bound cotton through a carbodiimide reaction, resulting in more robust anti-bacterial activity of the enzyme when bonded to cotton fibers than in solution. Retention of biological activity of the lysozyme-conjugated cotton-based fabric was an impetus for covalent attachment of organophosphorus hydrolase enzymes to cotton wipes as a method of wound-surface removal and decontamination of organophosphorous nerve agents [406].

Carbohydrate conjugates of cotton cellulose have also been prepared in the form of cotton-based wound dressings using a modification of the esterification of cotton cellulose with polycarboxylic acids outlined in Section 6.3.3. When citric acid was applied to cotton with the monosaccharide fructose and glucose using the pad-dry-heat curing technology, an esterified cellulose-citrate-linked ester of the monosaccharide was formed [407]. This same approach was also applied to grafting alginate onto cotton wound dressings to create a combination cotton-alginate wound dressing with enhanced properties of absorbency and elasticity. The conjugates formed during these esterification reactions linking cotton cellulose alginate have been termed "algino-cellulose" [408]. The alginate–citrate finishes of cotton gauzes were applied in various formulations containing citric acid, sodium hypophosphite, and polyethylene glycol and confer gelation properties upon hydration of the fabric.

6.4   DEGRADATION

6.4.1   GENERAL

The prime position of cotton among textile fibers is due to its capacity to withstand chemical damage during processing. Nevertheless such damage can occur, under certain circumstances, in which a study of degradation is so important. It is necessary to understand degradation to be able to prevent it. The agents most likely to degrade cotton during processing or subsequent use are acids, alkalis, oxidizing agents, heat, radiations, and enzymes, all of which will be considered in this section. A recent exhaustive review of textile degradation in general contains 827 references, of course many to fibers other than cotton [409].

Native cotton is nearly pure cellulose; the approximately 6% of minor constituents are usually removed during preparation for wet processing. The chemistry of cotton is therefore the chemistry of cellulose, but the effects of chemical changes on its textile properties depend on its supramolecular structure as well as on the nature of the changes themselves. The term degradation has been used in more than one sense. Originally, it implied a loss of tensile strength sufficient to render a fiber or fabric unfit for use. When it became known that one of the main causes of such loss was the severance by hydrolysis of glycosidic bonds in the cellulose chain, it was applied to this process. When this reaction occurs, it yields glucose. However, from a chemical point of view, a complete degradation of cellulose would yield the substances from which it was originally photosynthesized, namely carbon dioxide and water. Thus partial degradation, which interests the textile chemist most, may include chain scission by the hydrolysis of glycosidic bonds and the partial oxidation of AGUs without chain scission. It might be supposed that the latter process could occur without loss of tensile strength, but this has never been observed. Usually chain scission accompanies oxidation, but loss of strength can sometimes be attributed to other causes, such as cross-linking or changes of supramolecular structure.

The mechanical properties of cotton are very sensitive to some types of chemical changes. Quite small changes may affect them profoundly, sometimes even reducing the cotton to powder. However, small variations in the constitution of a large linear macro-molecule are not easy to investigate by the usual methods of organic chemistry. However, during the period 1920–1960, several tests, originally devised for monitoring industrial production, were developed into quantitative measurements for the characterization of slightly degraded cellulose [410]. The terms hydrocellulose and oxycellulose were already in common use for the water-insoluble products of acid hydrolysis and oxidation, respectively [411–413]. To study a particular degrading reagent, a series of progressively modified materials was prepared and characterized by measuring such properties as content of acid groups, reducing power, fluidity in cuprammonium hydroxide (CUAM), and tensile strength, the latter two are related to the DP. Valuable insights into the nature of various types of degradation were gained and the methods themselves were studied and improved. Stoichiometric methods are now available for measuring carboxy, aldehyde, and ketone groups, as well as terminal hemiacetals. IR spectroscopy has sometimes been used for the determination of these groups, but its application is limited by the small amounts of substituents usually present. It has of course been found more useful in the study of pyrolysis, along with other instrumental methods [414]. In many laboratories, CUAM for DP measurements has been replaced by the more stable and conveniently handled CUEN, cadoxen (cadmium ethylenediamine) or the Fe(III)–tartrate complex, FeTNa (EWNN in the German literature) [134,415].

The DP results rendered by most of these methods are low and inaccurate, especially when the celluloses are alkali-sensitive, i.e., contain carbonyl groups bringing about chain cleavage in alkaline solutions or active carbonyls, which initiate the stepwise depolymerization of the cellulose chains (known as the peeling reaction) according to the carbonyl elimination mechanism of Isbell et al. [416–420]. The most accurate method for DP determination in all celluloses is the nitration method, in which the cellulose is nitrated in a solution of nitric and phosphoric acids and phosphorus pentoxide, and dissolved in butyl acetate [420–422].

6.4.2 OXIDATION

Stoichiometric methods for measuring total carboxyl content depend on the exchange between a cation in solution and the solid cellulose substrate in its free-acid form:

$$RcellCOOH + M^+ \rightleftharpoons RcellCOOM + H^+ \tag{R6.6}$$

The equilibrium in the above reaction must be established as far to the right as possible and the main difference between the numerous methods that have been proposed is the way in which this is achieved [423]. The simplest of all is to measure the fall in concentration of a standard sodium hydroxide solution in which a known mass of material has been steeped. The solution should contain enough sodium chloride to equalize the concentrations of alkali in the solid and liquid phases by suppressing the Donnan effect [424]. This method is only valid if the cellulose contains no carbonyl groups since they cause the production of extra carboxy groups in the presence of alkali. The calcium acetate method [425] employs a dilute solution of this salt buffered to achieve a final pH $\geq$ 6.5, the fall in calcium concentration is measured by ethylenediaminetetraacetic acid titration. Results may be spuriously low if any carboxyl groups have formed lactones, because these are stable at pH 6.5. Methylene blue absorption [426] has two advantages over other methods because of the very high affinity of the dye compared with other cations: it does not require the material to be in its free-acid form and a degree of precision is obtainable even with materials of low carboxyl content. Virtually complete exchange is assured by establishing the following conditions in the final solution: [dye] $= 10^{-4}$ M, [Na$^+$]/[dye] $\leq$ 4, pH $= 8$. The change in the concentration of dye in the bath is measured spectrophotometrically. The methylene blue absorption of purified cotton is about 0.5 mmol per 100 g, which, if it is assumed to arise from acidic end groups, gives a number-average DP of about 1200.

Lactones in acid-washed samples of modified cotton are not hydrolyzed in methylene blue solution. They may, however, be determined along with free carboxyl groups by steeping the material in a solution containing potassium iodide and iodate, sodium chloride, and an excess of sodium thiosulfate [427,428]. Hydrogen ions from the material liberate iodine according to the reaction

$$6H^+ + 5I^- + IO_3^- \rightarrow 3I_2 + 3H_2O \tag{R6.7}$$

The iodine does not appear but reacts immediately with the thiosulfate, the consumption of which is determined by titration with the standard iodine solution.

Uronic acid groups (carboxyl groups at C6 in the AGUs) can be estimated from the yield of carbon dioxide on boiling with 12% hydrochloric acid for 8 to 10 h [429–431].

Two types of carbonyl groups may be introduced into cotton during degradation: aldehyde groups (including hemiacetals) and ketone groups. The former have strong reducing powers, but the latter do not. Some analytical methods measure aldehyde groups only, and some total carbonyl. The earliest measure of aldehyde content was the copper number, which is defined as the mass in grams of Cu(II) reduced to Cu(I) by 100 g of dry material on boiling with an alkaline copper solution of specified composition for 3 h [432,433]. The method is empirical, the amount of copper reduced per aldehyde group depends on the position of the group in the chain molecule. For example, a hemiacetal group reduces about 22 atoms of copper [434], but the two aldehyde groups in a periodate oxycellulose (see below) reduce 1.6 and 8.9 atoms, respectively. Thus, the copper number is a very sensitive measure of hemiacetal end groups and the fact that it is close to zero for scoured cotton is strong evidence of the absence of such groups in this material. The best available method of aldehyde determination consists of measuring the increase in carboxyl content of a material when it is treated with chlorous acid at pH 3 [435]. The copper number should be reduced to nearly zero in the process. If it is not, a plot of copper number against carboxyl generated at various times must be extrapolated to zero copper number, but the results should be treated with caution [436].

Total carbonyl content can be measured by means of one of the well-known condensation reactions such as the formation of oximes, phenylhydrazones, or cyanohydrins. Some of the methods proposed have serious limitations, but the reaction with sodium cyanide is usually considered to be satisfactory [437,438]

$$R'COR'' + NaCN + H_2O \rightarrow R' - \overset{\overset{\displaystyle OH}{|}}{\underset{\underset{\displaystyle CN}{|}}{C}} - R'' + NaOH \qquad (R6.8)$$

The consumption of cyanide is accurately determined by argentometric titration. The method was found to apply both to oxidized starch [439] as well as to cellulose [418]. The method is accurate and recently the coefficient of variance and standard deviation for an oxidized cellulose containing 5.60 mmol per 100 g of ketone groups, were found to be 0.0046 and 0.0068, respectively. The corresponding values for the carboxyl groups by the methylene blue method of the same samples were 0.001 and 0.031 [440].

The cyanide method is presently the only method for the determination of ketone groups in the polymers and was highly instrumental in the chemical characterization of degraded and oxidized celluloses. The use of this method enabled the development of the first two systems for the preparation of keto-cellulose, namely by mild oxidation with aqueous bromine at low pH values at room temperature [441,442] and by mild oxidation with hydrogen peroxide at pH 10 and 80°C [420,443].

If the carbonyl content is very low, it is better to use ^{14}C-labeled cyanide and determine the radioactivity of the cyanohydrin [444].

6.4.3 ACID

The degradation of cotton by acids consists of the hydrolysis of glycosidic linkages in accordance with Scheme 6.1.

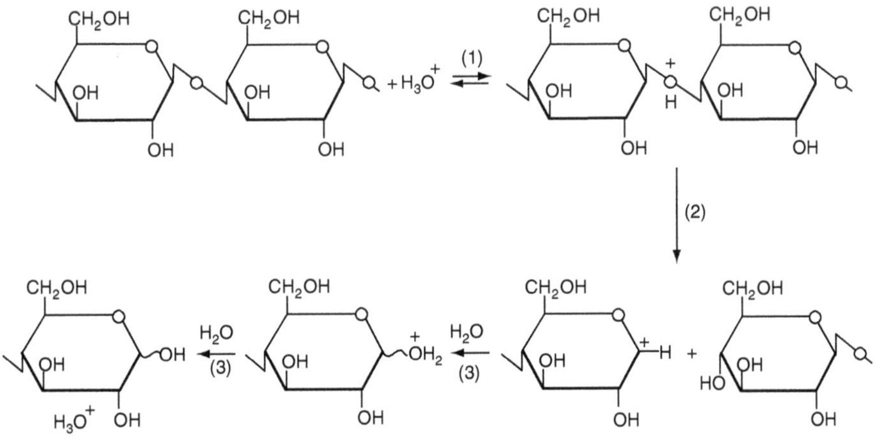

SCHEME 6.1

The reaction takes place in three steps, the formation of a carbonium ion as the glycosidic bond breaks (step 2) is the rate determining step. It is exactly analogous to the hydrolysis of simple glycosides [445]. In homogeneous hydrolysis, for example in 72% sulfuric acid, a yield of D-glucose as high as 90% can be obtained. In a textile context, only dilute aqueous acids come into contact with cotton, which on drying become concentrated, and hydrolysis is confined to the accessible regions of the fiber. Hydrocelluloses, materials of reduced DP having one reducing and one nonreducing end group per macromolecule, are formed. Their properties depend solely on the number and distribution of glycosidic linkages broken. This is governed by the concentration of hydrogen ions, the temperature, and the time of treatment; the nature of the anion is irrelevant. Hence, the tensile strength, copper number, and fluidity in CUAM of hydrocelluloses are uniquely related to one another irrespective of how they have been produced. The products of the early stages of hydrolysis are fibrous, but those with fluidities above about 44 rhe (1 rhe $= 10$ m^2/(N s)) are usually powders. After some time, the fluidity ceases to rise and a rate plot reaches a plateau. With native cotton, this occurs at a fluidity of 50 rhe, which corresponds to a DP of about 220. This is known as the leveling-off DP (LODP); with mercerized cotton, it is close to 180. The particles of an LODP hydrocellulose are generally identified with the crystallites in the original cotton.

Of the three stages in the acid hydrolysis of cotton, the first is very brief with a reaction rate 10,000 times greater than in the second stage. This is due to the presence of so-called weak bonds [446,447]. The nature of the weak bonds has been the subject of controversy, but it is now generally accepted that they are ordinary glycosidic bonds under abnormal physical stress arising during the original formation of the fibers [448,449]. The second stage represents the random hydrolysis of glycosidic bonds in the accessible regions of the fibers. The third stage consists of the end-wise attack of acid on the otherwise impervious crystallites. Thus, small soluble fragments are progressively removed, causing a continuous loss in weight but no significant fall of DP [450]. This is expected if there is an exponential distribution of crystallite lengths and all crystallites contain the same number of chain molecules [451]. Recent work has suggested that, while this may be true for mercerized cotton, it is only an approximation for native cotton [452,453].

When cotton is treated with very dilute solutions of hydrogen chloride in an aprotic solvent such as benzene it suffers severe degradation. This is because the small amount of hydrogen chloride in the solvent is redistributed in the water adsorbed on the cotton, forming a very concentrated aqueous solution of hydrochloric acid [454]. At low moisture contents, the sites of the consequent hydrolysis are near the ends of the cellulose chains. The relation between DP and copper number therefore differs from that for normal aqueous hydrolysis. However, as the moisture content of the cotton increases, the type of hydrocellulose produced approaches that found with aqueous systems.

6.4.4 ALKALI

Cotton is frequently treated with hot alkali in the processes preparatory to dyeing, as well as in the preparation of pure cellulose for research purposes. In the now nearly obsolete process of kier boiling as much as 4% of its cellulose content might be lost as soluble products. The observations that the loss in weight of hydrocelluloses in alkali boiling is directly proportional to the copper number led to the suggestion that short-chain materials were detached from the reducing ends of the hydrocellulose chain molecules and passed into the solution [455]. This was later fully confirmed and the mechanism of the process elucidated [456–464].

The basic mechanism operating in the hot alkaline treatments of cellulose is the beta-alkoxyl carbonyl elimination reaction of Isbell [416]:

$$\overset{\beta \quad \alpha}{R_1O-C-C-R_2} + OH^- \rightarrow [R_1O-C-C^--R_2] + H_2O \rightarrow R_1O^- + C=C-R_2 \qquad (R6.9)$$

According to this reaction any strongly negative group in the position β of an ether, where the α-carbon is carrying a hydrogen, will render the ether bond sensitive to alkali. The hydrogen on the α-carbon atom adjacent to a ketone or aldehyde group will become acidic enough to be removed by a base. This will be followed by an elimination of the alkoxyl group from the

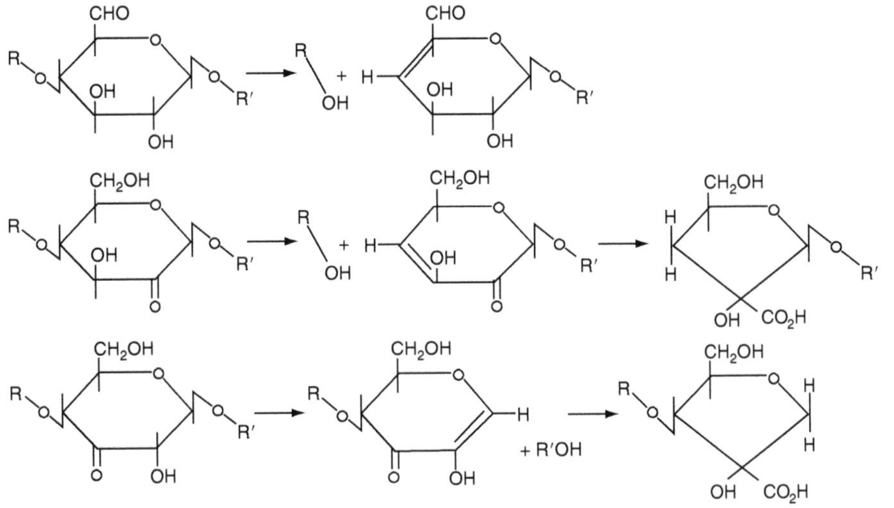

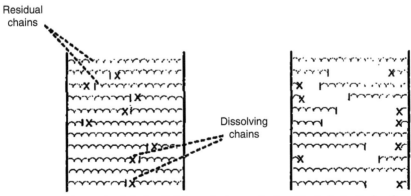

SCHEME 6.2

β-carbon atom so that an unsaturated product will be formed along with the cleavage of the etheric bond [465]. The upper equation in Scheme 6.2 illustrates the case when the negative carbonyl group is located on C-6. The double bond containing AGU was eliminated from the chain R—OH, which contains a reducing C-1 end. By the Lobry de Bruyn–Van Eckenstein transformation, the C-1 aldehyde is converted into a C-2 ketone, which will be in the position β to the ether bond and thus will cause another chain cleavage and elimination. This stepwise depolymerization, which is called the peeling reaction, continues until the whole length of the chain in the LORs will be depolymerized or until it reaches the crystalline region when a stopping reaction sets in. The 50% of the cleaved chains beginning with C-4 will remain unchanged (see Figure 6.4) [420,441,442,465,466].

When the carbonyl group is located on C-2 (see Scheme 6.2 bottom equation) the chain scission will occur on C-4, producing a new chain with a reducing end at C-1. This new chain will behave according to the upper equation in Scheme 6.2, which means that the depolymerization reaction will propagate along the chain and consequently the hot alkali solubility will be high.

FIGURE 6.4 (X) Reducing end of chain. (|) Nonreducing end of residual chain. Left: after oxidation. Right: after alkaline extraction and peeling.

With a C-3 ketone group (see equation at the bottom in Scheme 6.2), the scission will occur at C-1, producing a nonreducing chain and a diketo derivative, which may be transformed into a saccharinic acid. There will be no continuing depolymerization and the alkali solubility of such celluloses will be insignificant. The ketocelluloses produced by acidic bromine [420,441,442] and by alkaline peroxide [443], discussed above, behave in this manner. Aldehyde groups on C-2 and C-3, as obtained by periodate oxidation, will produce a rupture of the pyranose ring but no peeling reaction will ensue [442]. Hypochlorite oxidized celluloses may, depending on the pH of oxidation, contain carboxyl groups (pH > 10), aldehyde and ketone groups (pH 5–10).

The simplified Scheme 6.3 shows that the diketo moiety formed in the hot alkaline extracts of hydrocellulose can undergo the benzylic rearrangement and produce a colorless isosaccharinic acid, which dissolves in the extract and constitutes the major product of alkaline degradation. It is, however, not the only product. During the hot alkali treatment of bleached, hydrolyzed, or oxidized celluloses, a yellow discoloration is formed in the solutions, which obeys Beer's law [466]. The absorptivity D of the alkaline extracts is linearly related to the extent of oxidation or time of hydrolysis. It is also linearly related to the aldehyde group content and to 1/DP of the cellulose. This enables in many cases a simple and rapid determination of the DP of celluloses hydrolyzed and oxidized by several oxidizing agents (excluding hydrogen peroxide). Figure 6.5 illustrates this relationship for oxidized cellulose. A straight-line relationship also exists between the D values and the amount of the cellulose dissolved in the hot alkali (See Figure 6.6).

The ultraviolet spectra of the yellow chromophore obtained in alkaline extracts of a hydrocellulose (see Figure 6.7) is pH dependent and shows an isobestic point [467,468] at 270 nm, proving that the chromophore is a single molecule and not a mixture, and that it is stable in the whole pH range. A part of the chromophore appears in the visible range. The chromophore acts like an indicator and its absorbance decreases upon acidification, and darkens at higher pH values, which explains the well-known brightening effect of kier boiled fabrics upon acidification (the souring step) as well in other bleached cottons. The chromophore appears to be formed from the aldehydo-ketone or diketone intermediates (see Scheme 6.3). Its formation competes with the formation of isosaccharinic acid. The ratio of the rates of these two reactions is influenced by the presence of cations, their nature, and valencies. Most effective in the decrease of the chromophore concentration is the calcium ion [469,470]. The

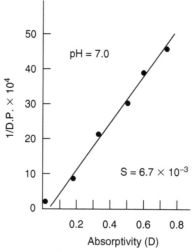

FIGURE 6.5 $1/DP$ versus D for cotton oxidized with hypochlorite at pH 7.

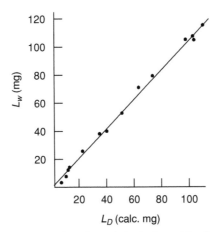

FIGURE 6.6 Gravimetrically determined loss in weight (L_w) on bicarbonate extraction versus loss in weight calculated from the D values on the basis on D of glucose, L_D. Pima cotton, hydrolyzed with 5N HCl at 25°C.

peeling reaction takes place in substituted and cross-linked cottons, as determined by measuring the yellowing of alkaline extracts of such celluloses. A schematic representation is shown in Scheme 6.4 [442].

The processes discussed above pertain to the main reactions of cotton with alkaline solutions up to temperatures of 120–130°C. Above 170°C, a random scission of glycosidic bonds in accessible regions occurs, leading to the rapid production of shorter chain molecules with new reducing end units. These immediately participate in the peeling and stopping reactions just described.

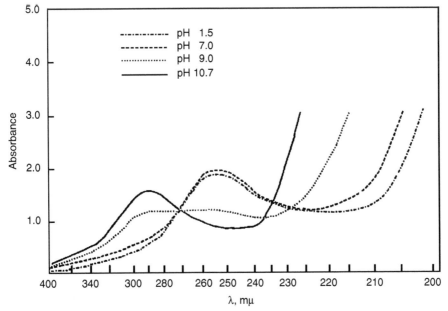

FIGURE 6.7 Ultraviolet (UV) spectra of hydrocellulose, boiled for 1 h in 5% bicarbonate at several pH values. Hydrocellulose prepared from Deltapine cotton fiber by shaking in 5N HCL for 72 h at 20°C. Amount dissolved in bicarbonate corresponds approximately to 3.7 mmol anhydroglucose units (AGUs).

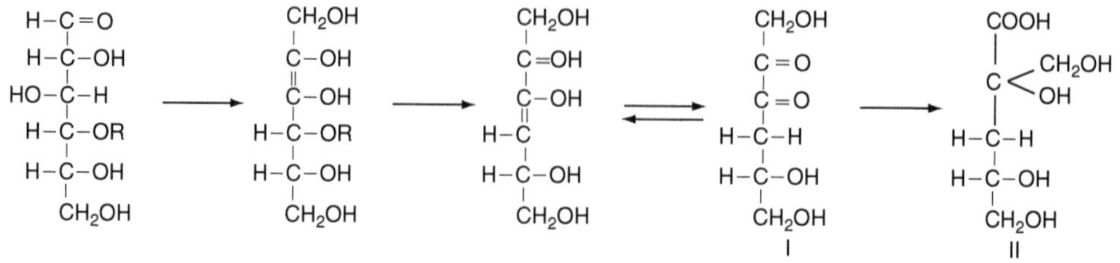

SCHEME 6.3

6.4.5 BIODEGRADATION

The biodegradation of cellulose is caused by enzymes known as cellulases [471–475]. Cellulases are produced by many microorganisms (bacteria and fungi). The most widely studied cellulases are of fungal origin, e.g., *Trichoderma* [471,473]. The cellulose-digesting bacteria of the rumen are a complex anaerobic community [476,477].

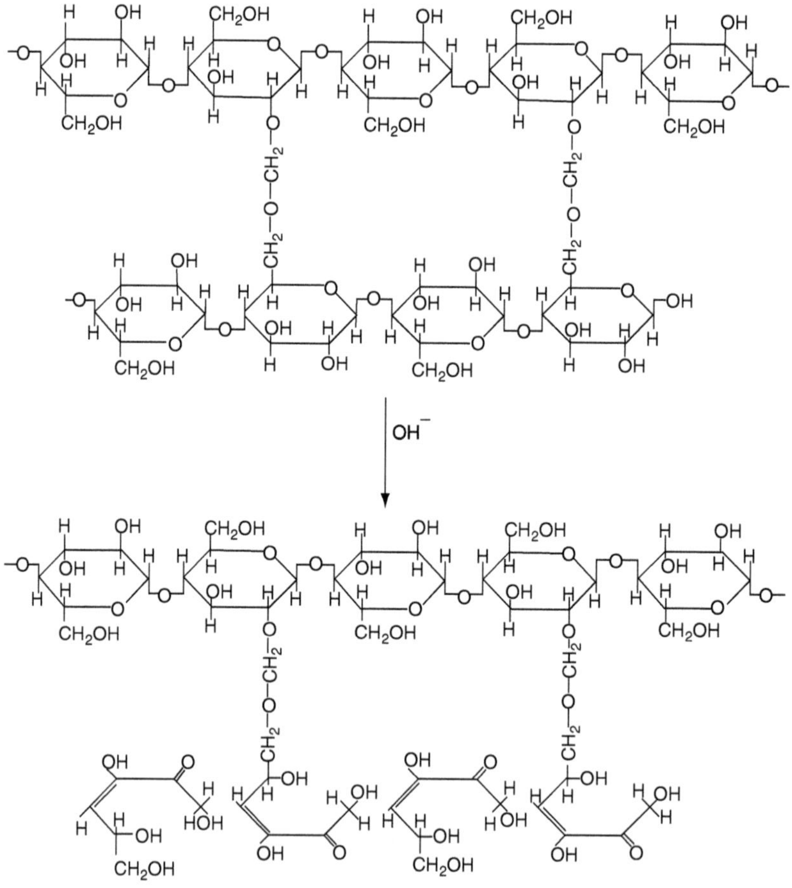

SCHEME 6.4

The primary reaction in the enzymatic degradation of the cellulose is hydrolysis, and degradation is a function of the available surface area and crystallinity of the cellulose. The biological degradation of the cellulose is discussed thoroughly by others [471,472]. The intentional use of enzymes in textile applications for desizing and finishing and some of the mechanisms of their actions is described in Section 6.6. Cotton fiber is a highly crystalline form of cellulose and the ability to degrade this form of cellulose is relatively rare among microbes, and, due to the solid nature and insolubility of the substrate, is not very rapid. The cotton fiber, after ginning, is about 95% cellulose with the other 5% mainly wax, pectins, and proteins (see Chapter 3). These noncellulosic nutrients are important to the maintenance of the complex microbial communities to which extensive cellulose degradation is typically ascribed, but in cotton fiber, these noncellulosic nutrients are in too low amounts to be of much significance.

Microbial damage, caused either by bacteria or fungi, can occur after boll opening, prior to or after harvest, and during storage, and to cotton fabrics. For example, "cavitoma" is a term for cotton that has undergone biological deterioration, when the extent of deterioration is not sufficient to cause a lowering of grade [478,479]. Cavitoma cotton has reduced strength. The microorganisms responsible for microbial damage are always present, but moisture content of around 9% [479] (although microbial damage has been detected at lower levels) is needed for their growth. Not all fungi or bacteria are cellulolytic [480,481]. The predominant microbial community is associated with the cotton fiber from the soil during cultivation [482] and is primarily aerobic. Significant quantities of oxygen or methane, which if produced could lead to spontaneous vapor-phase combustion, are not produced during microbial degradation of the cotton fiber [483]. Biological formation of methane occurs in anaerobic environments in which organic matter undergoes decomposition, such as the rumen of cattle. The same microbes that decompose cellulose do not produce methane. In addition, heat generated during the biodegradation of cellulose does not exceed about 180°F (80°C), which is well below the lowest temperature required for the ignition of cotton (see Section 6.4.6) and the autoignition temperature of cellulose (about 750°F (400°C)). Therefore, contrary to what has been suggested by some, the raw cotton fiber, unless contaminated with 5–10% oil and wet cotton fiber, is not capable of spontaneous combustion.

Tests currently available to identify and quantify microbial damage are reviewed by Allen et al. [484]. These tests include CUAM fluidity tests, pH determinations, reducing sugar determinations, microscopic techniques, and staining methods.

The U.S. Federal Trade Commission (FTC) has requirements [485] for products that make environmental marketing claims of biodegradability. A product claim should be substantiated by competent and reliable scientific evidence that the entire product will completely breakdown and return to nature, i.e., decompose into elements found in nature within a reasonably short time after customary disposal. Claims should be qualified to the extent necessary to avoid consumer deception about the products' ability to degrade in the environment, where it is customarily disposed, and the rate and extent of degradation.

The U.S. EPA considers cellulosic materials clearly biodegradable [486] and lists acceptable tests for biodegradable materials:

1. ASTM Method G21-70 (1984a)
2. ASTM Method G22-76 (1984b) [487]

These are tests for determining resistance of materials to fungi and bacteria, respectively. Research by Moreau et al. [488], using ASTM method G21-70, indicates that 100% cotton fabrics after 14 days can be completely degraded. However, the rate of degradation of any product is subject to climatic conditions. The Organization for Economic Cooperation and Development (OECD), based in Paris, France, also has test guidelines adopted in 1992 [489] for screening materials for ready biodegradability in an aerobic aqueous medium (OECD 301A-F) and inherent biodegradability (tests which allow prolonged exposure of the test material to microorganisms; OECD 302 A-C and OECD 304 A).

6.4.6 PYROLYSIS OR THERMAL DEGRADATION

The response of the cotton fiber to heat is a function of temperature, time of heating, moisture content of the fiber and the relative humidity of the ambient atmosphere, presence or absence of oxygen in the ambient atmosphere, and presence or absence of any finish or other material that may catalyze or retard the degradative processes. Crystalline state and DP of the cotton cellulose also affect the course of thermal degradation, as does the physical condition of the fibers and method of heating (radiant heating, convection, or heated surface). Time, temperature, and content of additive catalytic materials are the major factors that affect the rate of degradation or pyrolysis.

Heating cotton fiber in air or vacuum at 110 to 120°C drives off adsorbed moisture. Cotton dried in this manner still contains traces of moisture. As long as temperature is maintained below 120°C, there is little apparent change in the fiber after heating. Actually, however, carboxyl and carbonyl contents increase slightly and solution viscosity (DP) decreases slightly, but there is little change in tensile strength and textile properties.

The losses in DP and tensile strength, observed when purified cotton fiber is heated in air to 150°C, are proportional to temperature and time. When the fiber is heated over the same range, the losses increase with atmospheric moisture content; lower losses occur when cotton is heated in inert atmospheres or vacuum. The effect of heating cotton fiber as above is to produce cotton hydrocellulose in the fiber; water, heat, and, to a lesser extent, atmospheric oxygen bring about the cleavage of the glucosidic linkages in the accessible regions (LORs).

When cotton fibers are heated above 140°C, the tensile strength and viscosity decrease while carbonyl and carboxyl contents increase. Distinct discoloration of the fiber is seen; first, a yellowing (scorching) and then a deepening to brown as thermal degradation escalates. In air, the oxidation of cotton cellulose occurs at these temperatures. Complete loss of tensile strength develops after a few hours at or above 200°C in the absence of oxygen. Above 200°C, cotton cellulose is at the verge of thermal decomposition, and depolymerization begins. Between 200 and 300°C, primary volatile decomposition evolves. The early work on cellulose pyrolysis products was reviewed by Shafizadeh in 1968 [490]. These products can be trapped and analyzed in vacuum pyrolysis. The further decomposition to secondary products occurs when the primary products are heated at ambient pressure. The secondary products separate into a gaseous phase and a distillate (liquid) phase. The most recent work on cellulose pyrolysis is directed at optimizing the distillate fraction for conversion of celluloses or lignocellulosics to renewable fuels. A char is the only remaining solid after heating cellulose above 400°C. The proportions of gas, distillate, and char depend on the conditions of heating and the amount of inorganic catalyst or flame retardant present in the original cotton cellulose. When pure cotton cellulose is heated in vacuum, the gaseous phase constitutes about 20% of the total pyrolysis products, the liquid phase about 65%, and the char about 15%. Differential thermal analysis (DTA), differential scanning colorimetry (DSC), thermo-gravimetric analysis (TGA), gas–liquid chromatography-Fourier transform infrared spectros-copy (GC-FTIR), and direct pyrolysis-mass spectrometry (Py-MS, also referred to as evolved gas analysis, EGA) are techniques that have played important roles in following the pyrolysis of cotton fiber.

Several kinetic studies of the pyrolysis of cellulose have been reported [491–495] and, possibly due to the complications arising from impurities, have not led to any agreement on the kinetic order or the mechanism. However, Chatterjee found experimentally [496] and showed [497] from the results of Lipska and Parkera [492] a linear relationship between the square root of the weight loss versus time. This relationship was explained based on a two-step reaction sequence: initiation (i.e., random scission of glucosidic bonds producing reactive molecules) and propagation (i.e., decomposition with weight loss of the reactive molecules). This square root relationship was confirmed by Lewin and coworkers [498–500], who found, upon isothermal pyrolysis (at 251°C) of a series of carefully purified cottons and other celluloses of various origins, that the rate increases with decreasing crystallinity and increasing orientation and is inversely proportional to the square root of the DP (correlation factor of 0.940). The energy of activation increases linearly with crystallinity and extrapolated values of 30 and 65 kcal/mol are obtained for 100 and 0% LORs, respectively [498].

The isothermal pyrolysis in the presence of air proceeds at a much faster rate and higher weight losses are obtained as compared to vacuum pyrolysis at the same temperature. The first order rate constant obtained is linearly related to the expression: $[\%LOR + \sigma$ (% crystallinity)$]/f_0$ with a degree of correlation $r = 0.923$, where σ is the accessible surface fraction of the crystalline regions according to Tyler and Wooding [501], and f is the orientation factor. No correlation could be found with DP due to very rapid depolymerization. The fact that the rate is inversely proportional to the orientation and that it decreases with the increase in the thickness of the fibers indicates that the rate of the diffusion of the oxygen into the fibers controls the kinetics and that oxidation is the predominant process in air pyrolysis.

Pure and flame-retardant cotton fibers were studied by TGA and Py-MS [502] to measure the pyrolysis kinetics and primary pyrolysis products in order to propose mechanisms of cellulose degradative processes in the absence of oxidative processes or further degradation of the primary products. At a heating rate of 5°C/min, cellulose pyrolysis occurs in three consecutive stages, each with its own kinetic and product profiles [503]. The first stage, which occurred at 250–290°C, was characterized by a low energy of activation (114 kJ/mol) and a large negative entropy of activation (-137 J/(mol °C)). The volatile products from this stage consisted of only carbon dioxide, carbon monoxide, and water and accounted for only about 2% of the dry weight of cellulose. It was proposed that this stage consists of a random chain scission in the LORs of the cellulose followed by relaxation of the broken chains and dehydration, decarboxylation, or decarbonylation of AGUs previously affected by oxidization in the preparation of cotton. The chain scission was proposed to occur by a conversion of an AGU in a LOR from the normal $^4C_1$ to the $^1C_4$ conformation followed by a rate-determining attack of O6 on the acetal carbon (C1) of the affected AGU to leave a cellulose chain terminated in a 1,6-anhydro-AGU and another chain with a free C_4OH at the non-reducing end. This mechanism would account for the observed kinetic parameters and the strength loss observed at or below this temperature range.

The second pyrolysis stage follows the first stage at 290–310°C and is characterized by a higher energy of activation (165 kJ/mol) and a less negative entropy of activation (-40J/(mol °C)). The volatile products of this stage account for about 5% of the cellulose weight and include anhydroglucoses [e.g., levoglucosan (1,6-anhydro-β-D-glucopyranose), 1,6-anhydro-β-D-glucofuranose, and 1,4:3,6-dianhydro-α-D-glucopyranose]. The mechanism of this stage was proposed to be characterized by cellulose chain unzipping processes in the LORs from the chain ends formed in the first stage. The levoglucosan is formed by a scission process in the penultimate AGU similar to that in the first stage, resulting in a free levoglucosan and another 1,6-anhydro-AGU at the end of the chain. The cellulose chain with the free C_4OH terminal AGU is proposed to unzip by converting from the $^4C_1$ to the $^1C_4$ conformation and rear-side attack of the C_4 to displace the C1—O—cellulose bond. This displacement may occur before or after a displacement of the C_6OH by the C_3OH oxygen. These processes give the observed anhydroglucofuranose and dianhydroglucose. These processes are expected to have the observed intermediate activation energy and less negative entropy of activation. The attribution of the second pyrolysis stage to the LORs is supported by the behavior of ball-milled cotton cellulose, which shows no crystallinity by x-ray diffraction. Ball-milled cotton shows the same first pyrolysis stage as unmodified cotton cellulose, but the second pyrolysis stage of ball-milled cotton covers a temperature range of 290–354°C and consumes 80% of the original weight. A LODP has also been reported in the pyrolysis of unmodified cotton at about the same DP, weight, and increased crystallinity as observed in the acidic hydrolysis of unmodified cotton cellulose.

The third stage in the vacuum pyrolysis of pure cotton cellulose follows the second stage at 310–350°C and is characterized by a still higher energy of activation (260 kJ/mol) and a positive entropy of activation (123 J/(mol °C)). This stage accounts for approximately 80% of the original cellulose weight. The volatile products include, in addition to those from the first and the second stages, products formed by dehydration of the anhydroglucoses: 5-hydroxy-methyl-2-furfural, 2-furyl hydroxymethyl ketone, and levoglucosenone (1,6-anhydro-3, 4-dideoxy-β-D-glycero-hex-3-enopyranos-2-ulose).

The higher energy of activation and the relatively large entropy of activation are consistent with the rate of this stage, determined by disruption of the crystal structure of the native cellulose. The higher crystallinity of the remaining cellulose, the LODP, and the absence of the third stage in decrystallized cotton are consistent with such a mechanism.

The x-ray diffraction studies show that there are no changes in the crystal structure of cotton heated in sealed tubes below 250°C. Heating to 250–270°C causes loss of crystallinity and on further heating (280–300°C) the cellulose pattern disappears. The lowest temperature for self-ignition in air is 266°C, with a heating time of 2.5 h. What ignites, however, is not the cotton fiber but an organic residue, as the fiber has lost 60–70% of its original weight. Changes in the molecular structure can be followed readily by IR as the fiber is heated. The changes observed in the IR spectrum arise from the transformation of cellulosic hydroxyl groups into carbonyl and carboxyl groups, reduction in hydrogen bonding, breaking of glucosidic linkages, etc. As the fiber is heated to higher and higher temperatures, the characteristic cellulose bands disappear.

The practical aspects of the pyrolysis of cotton fibers are in the flame retardant cottons and in the ironing of cotton fabric. The subject of flame-retardant cotton fabrics and the means of producing them, which is discussed in Section 6.2.4, will only be touched upon lightly; extensive treatment of the subject can be found elsewhere [361]. Flaming combustion is a gas-phase oxidation of volatile fuels. These fuels are produced by pyrolysis of the solid fuel (cotton cellulose) as a result of the heat from the gas-phase combustion. Phosphorus-based flame retardants used in durable finishes for cotton textiles act by reducing the amount of volatile fuels formed in the pyrolysis, reducing the pyrolysis temperature, and increasing the amount of carbon-containing char remaining after the pyrolysis step. The durable phosphorus-containing flame retardants have specific effects on the kinetics and pyrolysis products of each stage of the pyrolysis [504]. Halogen-containing finishes act by generating volatile halides that act as free-radical traps and interrupt the gas-phase oxidation of volatile fuels. The phosphorus-containing, durable flame retardants for cotton textiles also suppress afterglow, which is the glowing combustion (a direct gas–solid oxidation) of the char remaining after pyrolysis or flaming combustion of cotton fabrics. A number of inorganic salts, such as potassium carbonate or magnesium chloride [505] suppress flaming combustion of cotton textiles, but they are not durable to washing and do not prevent afterglow in the residual char from heating the textile. Other inorganic compounds, such as ammonium phosphates, prevent both flaming and glowing combustion but are not durable to washing. Borax, boric acid, and other compounds of boron dried into cotton fibers inhibit combustion, but are not durable to washing.

Smoldering combustion can occur by sparks or lighted cigarettes falling on unprotected mattresses or cushions, and can lead to catastrophic flaming combustion involving furniture, etc. Smoldering propensity of cotton (e.g., upholstery fabrics and batting) has been found to be induced by alkali metal ions such as potassium, sodium, or calcium, as well as salts of iron, chromium, and lead [351,352]. Such ions are found naturally in raw cotton (which is used for batting and many upholstery fabrics) or as residues of dye assistants, softeners, detergents, etc. These compounds can effect an increase in the yield and reactivity of char formed during the pyrolysis of cellulosics [351]. Chemicals that inhibit the smolder of cellulosics appear to be compounds containing boron or phosphorus (e.g., boric acid and ammonium phosphates). These smolder inhibitors appear to intervene chemically in the oxidation reactions on char surfaces [351]. Boric acid, deposited on cotton batting fibers by way of the volatile methyl borate, is useful for preventing smoldering combustion in cotton batting. Other agents that are effective for cotton fabrics are not effective in cotton batting for preventing smoldering combustion, which is an oxygen-limited process controlled by the rate of diffusion of air through the batting to the fire.

Scorching of the cotton fabric begins in the ironing of cotton when flatiron temperatures reach about 250°C. Wet fabric scorches very much faster than dry fabric and, peculiarly, ease of scorching varies with cotton variety. The temperature of iron, the pressure of ironing, and the time of heating during ironing all have a relationship on the appearance of fabric scorching. Except for the effect of pressure, the effects of degree of temperature and time of heating are, as would be expected, that the degradation occurs with low temperature and long periods of heating or high temperature and shorter periods. In ironing, cotton fabric is in contact with the heated surface for only a short length of time, usually not more than 1 min. The appearance of scorching is a visual indicator that thermal degradation is occurring and can be taken as an end point for setting limits on ironing conditions. Tests have shown that for constant heating time the scorch temperature is reduced when ironing pressure is increased from 1.3 to 4 lb. When pressure is held constant and heating time is lengthened from 2.4 to 30 s, scorch temperature is again reduced. Scorch temperature of the fabric for these conditions ranged from 255 to 332°C. When cotton is exposed to temperatures of this magnitude for several hours, severe thermal degradation results, as has already been pointed out. Some of the original durable-press finishes for cotton cellulose promoted scorching in cotton fabrics, which had been exposed to chlorine bleaches, but this is no longer a problem with modern durable-press finishes.

Recent work on the thermal degradation of cellulose has shifted away from flame retardancy as the regulatory climate has changed. Areas of current research on thermal degradation of cotton include effects of heating during processing. The Acala 4-42 cotton was found to crystallize upon heating in a continuous curing oven for 10 min at 180, 190, and 200°C [259]. This thermal crystallization proceeds by first-order rate kinetics with an energy of activation of 32.5 kcal/mol. The accessibility of cotton decreased from the original 32.2 to 21.7% at 180°C, to 10.9% at 190°C, and to 4.6% at 200°C. The corresponding decreases in moisture regain from the original 7.92% to 7.01, 6.40, and 5.28%, respectively, were obtained. Rushnak and Tanczos [506] also obtained a decrease in moisture regain and dyeability of cotton upon heating. Back and coworkers [507,508] suggested that the increase in crystallinity is caused mainly by cross-linking reactions between aldehyde groups or between hydroxyl groups of adjacent chains in the LORs, producing interchain ether linkages. A similar assumption was made to explain the initial rapid weight loss in the isothermal pyrolysis at 250–275°C [498]. In addition, hydrogen bonded water molecules are removed from the polymer by the heating and enable the formation of new strong hydrogen bonds between hydroxyls of adjacent chains. These considerations explain the decrease in moisture regain and the increase in wet strength of heated paper.

Heating conditions during drying of cotton lint at the gin have been studied [509] to determine effects of drying conditions on cotton fiber quality. Excessive heating causes discoloration, strength loss, and reduced spinning quality, possibly from effects of heat on the noncellulosic components of raw cotton fibers. There has been considerable work on pyrolysis of cellulose and lignocellulosic materials to generate gaseous or liquid fuels from biomaterials and by-products. Richards [510] has done considerable work on pyrolysis mechanisms and useful materials from forestry by-products.

6.4.7 VISIBLE, ULTRAVIOLET, AND HIGH-ENERGY RADIATION

Photochemistry involves the interaction of visible and ultraviolet light with cellulose whereas radiation chemistry involves its interaction with high-energy radiations, such as from ^{60}Co γ-radiation. Light promotes the deterioration of cellulosic products, particularly cotton fabrics, and certain dyes or other additives greatly accelerate this process known as photo-tendering. Ionizing radiation is used to sterilize medical and bioproducts many of which are cellulosic. Several authoritative reviews are available [511–514].

The regions of the electromagnetic spectrum that are of interest in photochemistry are the visible (400–800 nm), the near-ultraviolet (200–400 nm), and the far-ultraviolet (10–200 nm) regions. To effect a chemical change directly the radiation must be absorbed by cellulose. Pure cotton cellulose, a saturated compound, lacks the structural features required to absorb light in the visible region. Even absorption in the near ultraviolet remains uncertain. Light with wavelengths greater than 310 nm (Pyrex glass filter) is not able to photolyze cellulose directly. Mercury emission at predominantly 254 nm can cause photodegradation and produce free radicals. It is not clear what structural feature is responsible for the absorption of radiation [514]. Irradiated carbohydrates do form a species absorbing at 265 nm [515] that exhibits an autocatalytic influence. The direct photolysis of cellulose with 254 nm radiation results in degradation and is independent of the presence of oxygen [516,517]. Changes are increase in the solubility, the reducing power, the formation of carboxyl groups, and lowering in the DP.

Although light of wavelengths greater than 310 nm cannot degrade cellulose directly, some other compounds such as dyes and some metallic oxides can absorb near-ultraviolet radiation or visible light and in their excited states can induce the degradation of cellulose. These reactions are designated photosensitized degradation but do not have a common mechanism. The photochemical tendering of cotton fabrics containing vat dyes has been recognized for many years [518–520]. Emphasis has been placed on practical problems involving textile performance rather than on mechanistic studies. It is known that this sensitized deterioration depends on the presence of oxygen in contrast to direct photolysis. The cotton cellulose is also subjected to degradation upon exposure to high-energy radiation. The radiation sources have included cobalt-60, fission products in spent fuel rods, and high-voltage electrons generated by linear acceleration [521]. High-energy radiation causes oxidative depolymerization of cellulose in the presence or absence of oxygen to yield hydrogen, one-carbon gaseous products, and multicarbon residues. At low doses, fibrous properties are retained. The effects of radiation dosages ranging up to 833 kGy on some of the physical and chemical properties of cellulose are shown in Table 6.1. The oxidative depolymerization of cellulose is shown by the increase in carbonyl and carboxyl groups and by the solubility of the irradiated residues.

The high-energy radiation forms macrocellulosic radicals that are stable in the crystalline areas of cellulose. These radicals can initiate reactions with vinyl monomers to yield grafted polyvinyl-cellulosic fibers with desired properties [522–524].

TABLE 6.1
γ-Irradiation Effect on Some of the Physical and Chemical Properties of Cotton Cellulose[a]

Dosage (kGy)	DP of Cellulose[b]	Carbonyl Groups (mol = g)	Carboxyl Groups (mol = kg)
0.00	4400	0.00	0.002
0.83	3100	0.01	0.002
8.3	880	0.05	0.005
42	320	0.22	0.015
83	210	0.36	0.023
417	79	1.30	0.070
833	56	2.66	0.139

[a]Purified cotton fibers irradiated in nitrogen at 298 K.
[b]Precision of these determinations is low.

6.5 WEATHER RESISTANCE

Most of the cotton fabrics, although capable of maintaining their strengths for many years indoors in a dry environment, deteriorate readily when kept in an outdoor environment. This is primarily because of bacterial attack and sunlight's actinic degradation (see Section 6.4.5 and Section 6.4.7). Although bottom weight (i.e., 8.0 oz/yd^2 and greater) cotton fabric has been used for years for tentage and awning material, its poor resistance to the above-mentioned outdoor hazards has limited its usable life, especially in rainy, damp climates. An entire industry has grown up in an effort to protect cotton fabric intended for outdoor use. Many treatments with agents that repel or kill microorganisms and that shield cotton fabric from actinic degradation have been proposed over the last half century. Formulations containing inorganic salts of chromium, copper, zinc, mercury, iron, titanium, and zirconium, as well as a number of organic compounds of sulfur, chlorine, and nitrogen, have found use as antimicrobial and sunlight-resistant agents for cotton fabrics intended for use outdoors [525,526].

Although damage from moisture and mildew are insignificant in most indoor environments, sunlight damage can be quite severe, for products like cotton drapes. Those formulations containing chromium or titanium, or vat or pigment dyes in the blue or grey shades have been the most successful. If titanium is used, it is usually used as TiO$_2$ in its tetragonal (anatase) form and is usually applied to that side of the drape, which faces the sunlight. If the drape is to be laundered, the treated side must also be coated with a vinyl or acrylic film. Environmental considerations and concerns for heavy metal ions in processing plants have begun to limit the choices open to the finishers in the United States.

For bacterial and mildew resistance, inorganic materials are applied in such a way as to permit them to be slowly released from some soluble form, thereby providing a constant minimal presence of the toxic agent over a long period of time. Agents such as chromium oxide (as in the pearl grey finish) [527,528], copper and zinc naphthenate, copper-8-hydro-xyquinolate, organomercury compounds, mixed agents containing both zirconium and chromium (i.e., the "Zirchrome" finish) [527,528], iron salts, and numerous organic compounds such as thiazolylbenzimidazole, octyl isothiazolinone, chlorinated phenols (e.g., pentachlor-ophenyl laureate), and the quaternary naphthenates have historically been employed by cotton fabric finishers to provide rot resistance. An excellent study comparing the effectiveness of a number of the commercially available fungicides and sunlight protectors over a 3-year period was reported in 1978 [526]. This study compared both inorganic and organic materials with and without a protective coating at several weathering sites. Of all the agents evaluated, those containing chromium salts or octyl isothiazolinone resisted the mildew and retained the strength best. Coated treatments always outperformed uncoated ones [526]. Many of the inorganic agents resulted in the deposition of insoluble inorganic pigments on and into the fabric yarns and interstices. These materials are frequently opaque, and although all of the above-mentioned chemical treatments were used primarily for bacterial and algae protection, it was understood that some of them would also provide a substantial resistance to sunlight degradation. This is especially true of the chromium, the titanium, and to a lesser extent, the zirconium compounds. For more extended usable life outdoors, these fabrics were often treated with additional fungicides and either a water-repelling wax or a vinyl or acrylic coating.

6.6   ENZYMATIC MODIFICATION

Enzymes are used in textile applications [529,530] for the hydrolysis of starch after desizing of cotton, for the removal of surface fibers, for giving cotton fabric a softer hand, and for enhancing the color of the dyed fabric. What differentiates enzymes from other catalytic systems is their remarkable specificity. As an example, starch polysaccharide is hydrolyzed by α- or β-amylase at the α-$(1 \rightarrow 4)$-glucoside linkage, whereas a cellulase has no effect at this site. This specificity is attributed to the active-site geometries of these proteins that are dictated by the distinctive three-dimensional shapes in aqueous solution. These complex three-dimensional configurations comprise alpha-helix designs, beta-pleated sheets, and hairpin turns as part of the unit structure. The enzymes that hydrolyze cellulose are known as cellulases or 1,4-β-glucanases [472]. Endoglucanases attack in the middle of the chain whereas exoglucases, also called cellobiohydrolases, remove cellobiose units of the chain end. The resulting cellobiose units are hydrolyzed to glucose by cellobiose hydrolases. The molecular weights of these proteins range from approximately 10,000 to 80,000 Da or more in the fermented broth. They are obtained from both bacterial and fungal sources. The three-dimensional structures of many of these proteins have been determined by x-ray crystallography. Enzymes that must attack insoluble substrates such as crystalline cellulose usually contain a domain that binds tightly to the cellulose and a catalytic domain attached to it by a flexible peptide. Hydrolysis of the glycosidic bond of cellulose (Figure 6.8) can occur with either retention or inversion of the anomeric configuration depending on the cellulase.

The endo-1,4-β-glucanases, exo-1,4-β-glucanases (cellobiohydrolases), and cellobiose hydrolases work synergistically. The catalytic sites of the endo-1,4-β-glucanases are generally located in a cleft whereas those of the exo-1,4-β-glucanases are in a tunnel-like structure. Examples of endoglucanase [531] and exoglucanase [532] structures are given in Figure 6.9.

The cleft in the endoglucanase permits it to attach to the crystalline microfibril (see Figure 5.3) and effects hydrolysis of a glucosidic bond within the polymer chain, creating new terminal sites. The polymer chain is then passed through the tunnel containing the active

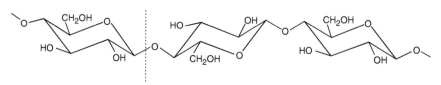

FIGURE 6.8 Glucosidic bond in the cellulose polymer that is hydrolyzed by cellulase (indicated by broken line).

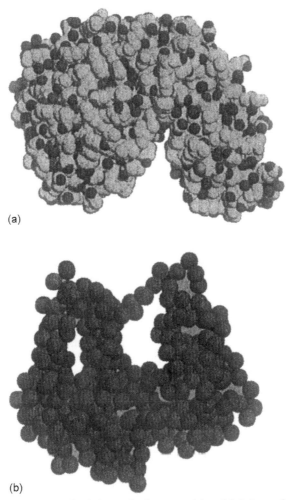

(a)

(b)

FIGURE 6.9 Molecular structures of a 1,4-β-endoglucanase (a) and 1,4-β-exoglucanase (b) showing the cleft and tunnel features, respectively. These drawings were prepared from files downloaded from the Brookhaven Protein Data Bank (http://pdb.pdb.bnl.gov). (a) (1CEC) and (b) (3CBH, Cα skeleton) were rotated to reveal the cleft and tunnel, respectively, with RasMol, Version 2.5, by Roger Sayle, freeware available from the Brookhaven National Laboratory web site.

site of the exo-1,4-β-glucanase (cellobiohydrolase) and the disaccharide cellobiose is liberated. The cellobiose is hydrolyzed to glucose by a cellobiose hydrolase. After the polymer chain is fragmented by attack of the endoglucanase, the smaller chain segments are rapidly removed from the microfibril by the exoglucanase. The microfibril is swiftly planed by the enzymes, leaving behind a crystalline material structurally similar to the original. The subjects of the structures and mechanisms of glycosyl hydrolases, such as cellulases, have recently been reviewed by Davies and Henrissat [533]. Modification of the cellulase structures and the composition of the mixtures are pursued as a means of tailoring their activity to better meet the practical objectives of the cotton textile industries.

It has been reported that crystallinity and accessibility to moisture do not change after enzymatic hydrolysis. Techniques employing light microscopy and staining of fibers with Congo Red can be used to determine the location of enzymatic attack for cotton and linen, but not for viscose rayon or ramie. In addition, SEM is useful for detecting changes in fiber morphology for all cellulosic fibers [534]. An important aspect of the hydrolysis of cellulose by cellulase is that the hydrolysis reaches a point where the hydrolytic activity levels off. This is reported to be a function of particle size, hydratability, and porosity [535].

The stability of enzymes is dependent on the hydrogen ion concentration of the environment. Cellulases are most stable at their isoelectric point (the sum of the cationic charges is equal to the sum of the anionic charges), but this is not necessarily the point at which catalytic activity of the particular enzyme is at its maximum. The greatest activity may be at higher or lower pH levels. At higher pH values enzymes ionize as weak acids and are precipitated by cationic agents, while at lower pH values they ionize as weak bases and are precipitated by anionic agents [536].

One of the main reasons for using enzymes instead of other chemicals as finishing agents for cotton cellulose is that they are environmentally safer. The small catalytic quantities will eventually degrade in the manner of proteins in general. With a typical formulation, cellulose is hydrolyzed with 0.2–0.4 g/l of the enzyme solution at a buffered pH of 4.5–5 and at a temperature in the range of 45–60°C for approximately 60 min or longer. Other enzymes designed to operate at higher pH values under neutral or basic conditions are also useful. Generally, the rate of a chemical reaction doubles for each 10° rise in temperature, but in the case of enzymes, there is a point above which most enzymes cannot be heated without loss of catalytic activity. Most common cellulases are rendered inactive or denatured by heating to approximately 65°C and higher. This inactivity is associated with unfolding of the three-dimensional structure. In contrast, cellulose is hydrolyzed much more rapidly under harsher acidic conditions by heating at 80°C in the presence of 2.5 N hydrochloric acid.

One of the major uses of cellulase on cotton is for removing fibers from the surface of dyed textile products to enhance color. This operation is referred to as biopolishing, and cellulase preparations have been incorporated in laundry detergents to eliminate or reduce surface fuzziness on cotton goods during the wash cycle. Enzymes are used also for biostoning of dyed garments, as done in the garment dyeing industry. Pumice stones have been used in recent years to soften garments and to remove color and produce unique washed-down appearances through abrasive action. However, such processes are not only environmentally unsound because of the disposal problem of ground-down stone particles after processing, but the stones are also very damaging to the processing equipment as well. The use of enzymes eliminates or substantially reduces the need for treatment with pumice stones [537,538]. The effectiveness of any of these operations for the removal of the material from the substrate is dependent on the type of mechanical action during processing. This includes the abrasive action of fabric-to-fabric contact or the cascading effect of aqueous solution on the cellulosic substrate [539].

The effects of cellulase enzymes on cotton substrates that were dyed with various classes of dyes have been reported. It was found that vat dyes do not inhibit cellulase activity on cotton, and hydrolysis of the substrate with subsequent color removal transpires. In contrast, cellulase activity on cotton was inhibited by the presence of some direct and reactive dyes. There is the probability that a dye–enzyme complex is formed with less activity than that of the free enzyme. The increased weight loss from enzymatic treatment on untreated cellulose is associated with a slight reduction in dye sorption, and this may be because of a reduction of amorphous regions in cellulose where dye molecules are usually sorbed. Cellulase activity is more prominent on mercerized cotton than on unmercerized cotton. This is because mercerization is associated with a decrease in the crystallinity of the cellulosic structure [540]. One major negative aspect of cellulase treatment of cotton fabric is the strength loss associated with the approximately 3–5% weight reduction under normal treating conditions. Of course, this is to be expected after cellulose is hydrolyzed. Future research efforts may lead to processes in which enzymes would be useful for total fabric preparation including not only the desizing of cotton, but also scouring and bleaching of cotton, as well as for the elimination of imperfections on the cellulosic substrate. The possibility also exists for the use of enzymes as catalysts for chemical modification of cellulose by esterification and etherification reactions.

6.7 CORONA

The phenomenon of *corona discharge* is defined as a nondisruptive discharge of the air between two electrodes separated by a small space and maintained at high electrical potential with respect to each other. The discharge creates a plasma (a partially ionized gas) containing numerous tiny thermal sparks that can alter the surface properties of materials placed in the discharge zone. A considerable amount of research was devoted to study the effects of corona treating wool and mohair fibers with both air and air–chlorine mixtures to improve the strength, shrink-proofing, and processibility of these fibers [541]. Subsequently these treatments were applied to cotton fibers with increases in fiber friction, yarn strength, abrasion resistance, and spinnability [542]. The corona treatment apparatus for cotton, basically developed by Thorsen, consisted of a sliver drafting system that converted six slivers into a thin, 18-cm-wide web that is conveyed between parallel glass-covered electrode plates separated by approximately 1 cm [543]. The treatment zone was maintained at 95°C with 15 kV at a frequency of 2070 Hz applied across the electrodes. Thorsen conducted a rather definitive study with cotton demonstrating that the treatment did not alter the basic fiber strength but did increase yarn strength by the order of 25%. He further demonstrated that the treatment had many potential practical applications including:

1. Improved mercerization with measurable increases in conversion to cellulose II
2. Better yarn stretch and setting properties
3. Increased dyeability
4. Increased production of free radicals making possible the induction of enhanced polymerization reactions and treatments

Thibodeaux and Copeland [544] reported further enhancements of the corona treatment of cotton webs by introducing a higher-frequency, high-energy (3300 Hz, 4.2 kVA) power supply. The most significant finding of this research was that an approximate 50% increase in frequency of the corona discharge lowered the cell temperature requirements from 95 to 65°C for optimum treatment. The increase in fiber friction measured by cohesion test on roving, processed from treated sliver, indicated that drafting tenacity could be more than doubled and thus the twist necessary to be put into the roving for cohesion could be significantly reduced (by as much as 20%). Finally, the study demonstrated that ring-spun yarn could be produced with significantly lower twists than feasible with untreated cotton, with higher or equal strength than control yarns with higher twist. This translated into an increased production rate between 15 and 20%.

Further improvements in the corona processing of cotton were made by Australian researchers at CSIRO. The basic approach of treating drafted web in a wide cell was improved by developing a corona machine that treated whole sliver passing between three 5 cm-diameter rollers mounted circumferentially on the surface of a 15 cm-diameter roller and spaced 4 cm apart. A spacing (0.5 mm) is maintained between the rollers as the sliver pass through and high voltage is maintained between the large and small rollers, leading to a corona discharge in the nip zones. The experiments on treating cotton–viscose blended yarns [545] indicated that corona treatment increased sliver tenacity, cohesion, and wettability. For a given level of power, treatment effectiveness is proportional to residence time, and for constant residence time, effectiveness is proportional to power. Further studies [546] with the same apparatus, using cotton slivers, produced similar results for the effects of power and residence time and further indicated that the effect was enhanced by temperature, diminished by moisture content, and was independent of four different gases (Air, CO_2, N_2, and Ar) used in the reactor. These studies were then extended to study the effects of corona on yarn strength, fabric strength, and processing efficiency [547]. The strengths of yarns spun from corona-treated cotton slivers increased, especially for lower twists and coarser counts. Strength was also increased by repeated treatments. There was no difference in fabric strengths when comparing treated and untreated cottons, despite the fact that the treated cottons had been spun at lower twists. The hand of fabrics from treated cottons was somewhat harsher than fabrics from untreated cottons spun at the same twist level. However, this harshness virtually disappeared with fabrics from the lower twist yarns from the treated cottons.

6.8 DYEING

Cotton requires some pretreatment (referred to as preparation) prior to dyeing and printing [548]. The preparation processes of cotton textiles include singeing, desizing, scouring, bleaching, and mercerization. These treatments remove natural and human-induced impurities, i.e., noncellulosic constituents and other unwanted substances, and increase the affinity of cellulose for dyes and finishes. In addition, color enhancement can be accomplished through the treatment of cotton with cellulase enzymes either before or after dyeing. The increased enhancement occurs because of the removal of fibers, which give the surface of cotton fabric a frosted appearance [549].

The dyeing of cotton has been extensively reviewed. The composition of the dye molecule, the dyeing conditions, and the nature of the cellulosic substrate influence the kinetics and equilibria in dyeing cellulosic fibers. Cotton is dyeable in fiber form (raw stock), yarn, or fabric form with an extensive number of dye classes, including, azoic, direct, indigo, pigment, reactive, sulfur, and vat dyes [339,550–557]. The selection of the appropriate dyes and dyeing processes for cotton is based on numerous factors depending on the application for which the dyed cotton fibers are to be used. A detailed review of these considerations and the properties of the dyes are beyond the scope of this book.

Anionic dyes used on cotton are essentially water soluble because the dye molecule contains sulfonic acid groups. The most widely used dye classes for coloring cotton are the direct and reactive ones. The dyeing of cellulosic fibers with direct dyes is relatively simple, as the dyes are affixed by hydrogen bonding and van der Waals forces. More elaborate procedures are employed for dyeing cotton with sulfur, vat, and reactive dyes. Sulfur and vat dyes are generally water insoluble, but they are solubilized as part of the dyeing process by converting the dye to its leuco form as the soluble sodium salt. After diffusion of the dye into the fiber during the dyeing process, the water-insoluble keto form is formed again upon oxidation. Thus, the insoluble dye remains trapped within the cellulosic fibers.

Most dyes do not chemically react with the cellulose molecule to affix the color. The roles of interchain hydrogen bonding and van der Waals forces in the application of azoic, direct, sulfur, and vat dyes are the physical and chemical effects and are not classical chemical reactions. True chemical reaction between cellulose and the dye molecule occurs with reactive dyes, which comprise different chemical types (see Section 6.2.3). Such chemical reaction results in covalent bond formation between the dye molecules and the C6 hydroxyl groups of the cellulosic chains. In many cases, these covalent bonds are formed at temperatures ranging from approximately 25°C (over an extended period of time) to 80°C for 30–60 min at pH 11–12. The reactive dyes, such as di- or monochlorotriazine or other suitable derivatives, react with cellulosic hydroxyl groups by splitting out HCl and forming ether linkages (etherification). Sulfatoethylsulfone reactive dyes react with cotton cellulose hydroxyl groups in alkali solution forming ester bonds (esterification). Cottons dyed with these dyes have excellent colorfastness to washing. Of course, any laundering of the dyed cotton should be conducted without bleaching agent, such as sodium hypochlorite, which will destroy the chromophore and also cause cleavage of the covalent bond [558,559].

Cotton also can be colored with pigments, which are water-insoluble colorants of either inorganic or organic composition. As such, pigments have no affinity for cellulose and are used with a binder, such as a polymeric resin, that secures the colorant to the substrate. Their lack of water-solubilizing groups and solubility in the aqueous medium distinguishes pigments from dyes. Pigments are applied to textiles by padding, printing, or batch exhaustion application methods [560].

Normally, cotton is not dyeable after it has been cross-linked with N-methylolamide agents, such as DMDHEU. This is because the fibers are cross-linked in a collapsed state at elevated temperatures and they cannot swell adequately in aqueous solution to accommodate the relatively large dye molecules. However, cross-linked cotton is dyeable with anionic dyes under acidic pH conditions if reactive alkanolamines or hydroxyalkyl quaternary ammonium salts are incorporated in the finishing formulation. In addition, cotton cross linked with polycarboxylic acid, such as BTCA, or citric acid, is dyeable in similar fashion by using the same methods. Thus, these cross-linked cottons have affinity for acid, direct, and reactive anionic dyes at the pH range of 2.5–6.5, depending on the exact chemical composition of the substrate [561–564].

Basic dyes are not normally used to color cotton textiles, but they do react with the few carboxyl groups in the regular cotton or with the greater number of groups in the oxidized cotton (methylene blue for staining oxidized cotton). In addition, cotton that has been cross linked with polycarboxylic acid has affinity for basic dyes because of the presence of free carboxyl groups. However, the basic dyes are not strongly bonded to the substrate, the color is rather easily removed [565], and has poor lightfastness. All other dyeing processes produce dyed cotton of varying degrees of fastness, to washing and to light. Color and brightness of shade vary with the dye class and dyeing process, as does the cost of dyeing.

7 Physical Properties of Cotton

CONTENTS

7.1 MATURITY AND FINENESS

The term fiber maturity is generally understood to refer to the degree of development or thickening of the fiber secondary wall [566,567]. Fiber maturity is a function of the growing conditions that can control the rate of wall development and of catastrophic occurrences such as premature termination of growth due to such factors as insect infestation, disease, or frost. As we have seen, the fiber develops as a cylindrical cell with a thickened wall. As the diameter of the fiber cylinder is largely genetic or species-dependent, a simple absolute measure of the thickness of the fiber secondary wall is not sufficient to define maturity. Probably the best definition of cotton fiber maturity has been proposed by Raes and Verschraege [568] who state "...the maturity of cotton fibers consists in defining it as the average *relative* wall thickness." What is implied in this statement is that maturity is the thickness of the cell wall relative to the diameter or perimeter of the fiber.

A term that has more recently gained popularity is the degree of thickening, Θ, defined as the ratio of the area of the cell wall to the area of a circle having the same perimeter as the fiber cross section, or $\Theta = 4\pi A/P^2$ where A is the wall area (μm^2) and P the fiber perimeter (μm).

In selecting a cotton to be used in a manufacturing process, it is important to have a knowledge of the maturity of cotton, which will determine the ultimate quality of the product as related to dyeability and ease of processing. Immature cottons tend to not dye uniformly, and result in large processing wastes in large numbers of spinning and weaving breaks and faults.

Fiber maturity can be measured directly or indirectly. In general, the direct methods are more accurate and precise, but are much slower and more tedious than the indirect methods. In practice, direct methods are used to calibrate or standardize the indirect methods.

Three of the most significant direct methods include:

1. The caustic swelling [569] test, in which whole fibers are swollen in 18% caustic soda (NaOH) and examined under the light microscope with a specific assessment of the relative width of the fiber versus its wall thickness used to identify a fiber as mature, immature, or dead.

2. The polarized light test [569–571], in which beards of parallel fiber are placed on microscope slides on a polarized light microscope using crossed polars and a selenite retardation plate. The interference color of the secondary wall will be a direct measure of its thickness and thus maturity. Generally, mature fibers appear orange to greenish yellow whereas immature fibers appear as blue-green to deep blue to purple.

3. The absolute reference method of *image analysis* of fiber bundle cross sections [572–574], wherein an image analysis computer system is used to automatically measure the area and perimeter of several hundred fiber sections and statistically analyzed to measure the average Θ and perimeter.

The indirect methods are characterized by the need to be rapid as well as accurate and be reliable enough to be used in the cotton marketing system. These methods may be divided into the double compression airflow approach and near infrared reflectance spectroscopy (NIRS). Examples of the former (airflow) method are the Shirley Developments Fineness and Maturity Tester (FMT) [575], and the Spinlab Arealometer [576]. The NIRS approaches have been developed and discussed by Ghosh [577] and Montalvo et al. [578].

The term fiber *fineness* has had many interpretations and understanding in fiber science. Some of the most important parameters used to define fineness include:

1. Perimeter
2. Diameter
3. Cross-sectional area
4. Mass per unit length
5. Specific fiber surface

Of all of these five parameters, the perimeter has proven to be the least variable with growing conditions and is essentially an invariant property with respect to genetic variety. For this reason perimeter has become recognized by many as inherent or intrinsic fiber fineness. Because of the irregularity of cotton fiber cross sections it is very difficult to measure a real diameter (this would presuppose that the fiber was circular in cross section). Similarly, cross-sectional area, mass per unit length, and specific fiber surface are dependent on maturity and thus are not real independent variables that we desire. However, from the standpoint of the spinner, the most important of the possible fineness parameters listed above is mass per unit length. A knowledge of this parameter allows the selection of fibers based on the minimum numbers of fibers required to spin a certain size yarn, i.e., the finer the yarn, the finer the fiber required.

Fiber fineness or mass per unit length can be measured both directly and indirectly. The direct method [579] consists of selecting five separate bundles from the sample. In each bundle or tuft, the fibers are combed straight and each is cut at the top and bottom to leave 1-cm-long bundles. The fibers from each bundle are laid out on a watch glass beneath a low magnification lens and 100 fibers are counted out from each of the bundles, compacted together, and weighed separately on a sensitive microbalance. For each bundle, we obtain the fiber fineness in 10^{-8} g/cm. The closest thing to an indirect method for measuring fineness is the micronaire test [580]. However, as we will show, the micronaire test actually measures the product of fineness and maturity. It is based upon measurement of airflow through a porous plug of cotton fibers. In the standard micronaire test, 50 gr (3.24 g) of fiber are loosely packed into a cylindrical holder. The cylinder and the walls enclosing it are both perforated to allow the flow of air under pressure that compresses the fiber into a 1-in. diameter by 1-in.-long porous plug that will offer resistance to the flow of air under 6 lb/in.^{2}. Research has shown that the flow through the cotton is given by $Q = aMH$ where a is a constant, M is maturity, and H is fineness. These results imply that for a constant maturity, a micronaire instrument will be nearly linearly dependent on fineness. However, for samples of various finenesses and maturity, it has been demonstrated that there is a quadratic relationship between the product of fineness and maturity (MH) and micronaire. This relationship is best expressed by the quadratic $MH = aX^2 + bX + c$, where X is micronaire and a, b, and c are constants [566]. Thus, given that any two of the parameters (fineness, maturity, or micronaire) are known, the third can be determined, and the processor will have a much more complete picture of the quality of the cotton being processed.

7.2 TENSILE STRENGTH

An accurate knowledge of the tensile behavior of textile fibers (their reaction to axial forces) is essential to select the proper fiber for specified textile end-use applications. However, to have meaningful comparisons between fibers, experience has shown that it is necessary to conduct measurements under known, controlled, and reproducible experimental conditions [581]. These include mechanical history, relative humidity and temperature of the surrounding air, test or breaking gauge length, rate of loading, and degree of impurities. Mechanical history is important because fiber could be annealed, extended beyond its elastic limit, or otherwise affected by mechanical manipulation. Cotton is the only significant textile fiber whose strength increases with humidity while most others are weakened by increased moisture. Textile fibers universally lose strength with increasing temperature. It seems logical that the longer the breaking or gauge length, there is more chance for an imperfection to occur that will cause a failure. Likewise, the presence of impurities in a material will tend to lead to disorder and weakness.

One definition of strength is the power to resist force. In the case of an engineering material such as textile fibers, this can be translated as a breaking strength or the force or load necessary to break a fiber under certain conditions of strength. Although textiles will be forced to endure a wide variety of forces and stresses, experience is that tensile breaking load is an excellent benchmark of fiber strength. The general approach to relative ranking of the strength of materials is with breaking stress, i.e., the tensile load necessary to break a material normalized for its cross-sectional area. This is expressed by the equation, $\sigma = T_b/A$ where T_b is the breaking tension, and A is the cross-sectional area. The CI units for σ are N/m^2 (or Pa) [315]. However, when dealing with textiles it is often more convenient to think of strength in terms of force per mass rather than per area. When dealing with fibers this translates into force per mass (m) per fiber length (l) that defines *specific stress* (tenacity) as given by $\sigma_{sp} = T_b/(m/l)$ (N m/kg or Pa m^3/kg). Because of the magnitude of quantities dealing with textiles, it has been found to be more convenient to go to the tex system for linear density (1 tex = 1 mg/m) and for single fibers to measure load in millinewtons with the resultant stress units of mN/tex or gf/tex. Here gf refers to grams force or grams weight, which is the force necessary to give 1 gram mass an acceleration of 980 cm/s^2.

The values for the tensile strength of single fibers range from about 13 gf/tex (127.5 mN/tex) to approximately 32 gf/tex (313.7 mN/tex) [582]. Calculations based purely upon bond strengths of the cellulose molecule would predict much higher strengths for cotton, but other factors also contribute significantly to determine ultimate fiber tenacity. These include crystallite orientation, degree of crystallinity, fiber maturity, fibrillar orientation, and other features of the fiber structure. Although single-fiber testing is quite tedious and time-consuming, considerable studies in the past show rather consistent findings including that:

1. Fiber breaking load increases with fiber coarseness, though not in direct proportion to fiber cross-section.

2. Breaking tenacity correlates well with fiber length and fineness.
3. There is a correlation between fiber breaking load and fiber weight within any single variety [583].

Although, as mentioned above, the procedures for measuring single fiber breaks have made this type of investigation nearly prohibitive, especially in the case of quality control testing, a new instrument has been designed that shows potential for making single-fiber testing more feasible.

The most significant advance in the technology of fiber strength measurements has been the development of the Mantis, a single-fiber tensile tester [584]. Mantis's unique design allows for rapid loading of single fibers between the breaker jaws without use of glue and has computer system to control single-fiber breaks and record both stress–strain curves and other pertinent data. The single fiber test used by the Mantis consists of two measurement modes: mechanical and optical. Fiber mounting is semiautomatic, i.e., the operator places a fiber across the jaw faces and the fiber is straightened by a transverse airflow caused by two lateral vacuum pipes. Small jaws clamp the fiber ends and a slight stress (<0.2 g) is applied to remove crimp. Optical measurements are accomplished by detection of the attenuation of infrared radiation by the presence of the fiber. The degree of attenuation is proportional to the fiber's projected profile (ribbon width). A uniform stress is applied, causing the elongation of the fiber until it breaks. A plot of grams force versus elongation is provided until the fiber breaks. In addition to the stress–strain curve, Mantis supplies the force to break, T_b (g), the fiber ribbon width, RW (μm), and the work of rupture (J).

The main purpose of testing raw cotton's strength is to predict the strength of yarn spun from the fiber. As yarn strength is determined by not only fiber strength but also by fiber to fiber interactions as induced by length, friction, and degree of twist, it has been found that breaking bundles of parallel fibers give a better predictor of yarn strength by simulating the combination of fiber tenacity and interaction. The two most commonly used bundle testers are the Pressley and the Stelometer testers [585]. The Pressley operates with a flat bundle of parallel fibers clamped between a set of jaws that may be operated such that there is essentially no gap (zero gauge) between the clamps holding the fibers or that there is a 0.125 in. spacing (eighth-inch gauge) between the clamps. The loading on the jaws is initiated by a weight rolling down an inclined plane. When the breaking load is reached, the bundle breaks, the jaws separate, and the breaking force is read from an attached scale. The force is determined by the length of travel of the carriage weight and is thus applied to the bundle specimen at a nearly constant rate of loading. The mass of the broken bundle is also measured and the breaking tenacity is calculated. There are, however, some mechanical problems with the Pressley and as a result, an alternative bundle tester has been developed that is almost universally accepted as the method of choice. The Stelometer operates on the principle of a pendulum and by proper adjustment can be set to operate at a constant rate of loading of 1 kg/s. The fiber bundle to be tested is mounted between Pressley clamps set to in. The clamps are mounted between a pendulum and beam that are set to allow the pendulum to rotate about a fulcrum. When the beam is released, it rotates about a point, causing the pendulum to rotate in such a fashion as to produce a constantly increasing rate of loading of the bundle under test.

Cotton bundle strengths range from a low of about 18 gf/tex (176.5 mN/tex) for short coarse Asian cottons to a high of approximately 44 gf/tex (431.5 mN/tex) for long fine Egyptian cottons. The degree of crystallite orientation, the fine-structural parameter obtained by x-ray diffraction, is directly related to the bundle strength [586]. This parameter is a measure of the angle of the fibrils spiraling around the fiber axis. The angle varies from about 25° for Egyptian cottons to about 45° for the coarser and weaker species. The degree of crystallite orientation increases with decreasing spiral angle. In general, the more highly oriented fibers are stronger and more rigid. In general, cotton strength increases with moisture content and decreases with temperature.

The reaction of a material to a tensile stress is to stretch or elongate. This is measured as *tensile strain*, defined as the elongation or increased length per initial length. Strain is a dimensionless unit that is usually expressed as a percent (percent elongation).

7.3 ELONGATION, ELASTICITY, STIFFNESS, RESILIENCE, TOUGHNESS, AND RIGIDITY

In the previous section, we discussed the evaluation of the strength of cotton under conditions of static or steady loading. In practice, textile fibers are exposed to a variety of dynamic forces and their response to "stress in motion" will better characterize their performance during processing or in end use. To quantify a material's dynamic response, it is necessary to record the elongation corresponding to increasing load [582]. A typical load–elongation or stress–strain curve is shown in Figure 7.1.

This represents the response of cotton loaded to the point of breaking. It begins with a curvilinear region (AB) in which fiber crimp or kinkiness is removed as load is applied. The crimp of a cotton fiber is a minor parameter compared with the other factors in the stress–strain properties of cotton, and consequently little quantitative consideration has been given to it [587]. However, it has long been recognized that the crimp of a fiber plays a major role, leading to the phenomenon of fiber cohesion, a property that causes materials to cling together. Without cohesion, it would literally be impossible to spin yarn from staple fibers. In the case of the synthetic fibers, crimp must be artificially induced by elaborate texturing schemes.

With further loading of the fiber, the curve (BC) becomes linear. In this region, stress is proportional to strain with the ratio referred to as the Hookean slope (i.e. follows Hook's law). The point C at which the curve becomes nonlinear is referred to as the yield point where the loaded elements begin to deform in a nonelastic or irreversible fashion and redistribute the

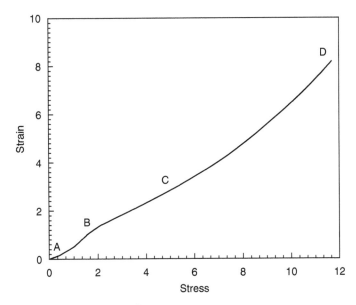

FIGURE 7.1 A typical stress–strain curve for cotton.

stresses. From here to the breaking point (D), the curve becomes essentially linear. The stress–strain curve is strongly influenced by any of several conditions as the rate of loading, sample moisture content, specimen length (where structural imperfections in the fiber come into effect), and extent of mechanical preconditioning.

The elongation of cotton is expressed as percent elongation taken at the point of breaking, hence the term elongation at break [581]. For most cotton, elongation at break, or just elongation, is in the range 6–9%. The effect of moisture is most pronounced on elongation. An elongation of about 5% at low relative humidity will increase to about 10% when the relative humidity is almost at the saturation point. The adsorption of water in the pores and amorphous regions of the fiber serves to reduce interfibrillar cohesion and to relieve internal fiber stresses. A more uniform distribution of applied stresses is thereby realized. The internal fibrillar orientation can be seen with x-ray diffraction. Swelling treatments, such as mercerization, ethylamine, and liquid ammonia treatments, affect the fiber elongation in a far greater fashion than water does.

Whereas the parameter elongation is the ability of the fiber to undergo deformation, elasticity is fiber's ability to return to its original shape when the loading is released. This property is highly time-dependent. Young's modulus, the ratio of stretching stress per unit cross-sectional area to elongation per unit length, may be calculated from data taken at either the beginning of elongation (initial modulus) or at the breaking point. Initial Young's moduli for cotton range from about 80 g/den for Sea Island down to 40 g/den for Asian cottons. The elasticity of cotton is imperfect because it does not return to its original length after stretching. When a fiber is stressed and allowed to recover, Young's modulus is found to be approximately one third of the initial value, indicating that some permanent deformation is now present in the fiber.

Elastic recovery of the fiber can be estimated by two methods. In one method, fibers are preconditioned mechanically by cyclic loading to specific levels. The recoveries after stretching and releasing to zero stress give elastic recovery, the ratio of recoverable elongation after a specified cycle to total elongation at the end of the cycle. In this method, elastic recovery is a function of percent elongation; elastic recovery drops curvilinearly from 0.9 at about 1% elongation to about 0.4 at the elongation at break (approximately 6%). In the second method, elastic recovery is resolved into three components: immediate elastic recovery, delayed elastic recovery (5 min after removal of load), and permanent set (or stretch). There is no mechanical preconditioning; a new sample is tested at each successive cycle to higher loadings. Percent elongation at each load is plotted against percent of total elongation to give a plot resembling a phase diagram.

The resilience of a fiber is the ratio of the energy absorbed to the energy recovered when the fiber is stretched and then released. To obtain this index, the areas under the stress–strain curves for the extension phase as well as the recovery phase are measured. The ratio of extension to recovery areas is the index of resilience. As would be expected cotton is not a very resilient fiber. Another parameter from the stress–strain curve is toughness or the energy to rupture. Toughness is determined from the area under the stress–strain curve measured up to the point of break. It may be closely approximated by the product of the breaking stress and the strain at break divided by 2. The units of toughness are therefore also gf/tex or mN/tex (CI). Values for cotton's toughness range from about 5 to 15 mN/tex. Compared to unswollen fibers, toughness is increased considerably by swelling treatments that end in drying without tension. However, swelling of fibers followed by drying under tension decreases toughness.

Rigidity of the fiber is another elastic parameter that is of great significance in describing a fiber's resistance to twisting. Thus, rigidity will obviously have applications to the spinning of textile fibers. Sometimes referred to as "torsional rigidity," it is defined as torque necessary to impart unit twist or unit angular deflection between the ends of a specimen of unit length [581].

In analogy to Young's modulus, the shear modulus or modulus of rigidity is defined as the ratio of shear stress to shear strain or as the ratio of torsional force per unit area to the angle of twist (displacement) produced by the torque. The finer varieties show less rigidity than coarser fiber: fine Egyptian cottons are in the range 1.0–3.0 mN/m^2; American cottons, 4.0–6.0 mN/m^2; and coarse Indian cottons, 7.0–11.0 mN/m^2. It may be convenient to introduce a specific torsional rigidity of unit linear density (tex) that is independent of fiber fineness. This may be defined as $R_t = \varepsilon n/\rho$, where R_t is the specific torsional rigidity (mN mm^2/tex^2), ε is a shape factor that equals approximately 0.7 for cotton, n is the shear modulus, and ρ is the density. Rigidity will vary with growth conditions and fiber maturity. Fiber rigidity increases with temperature and decreases with moisture content. Difficulties in fiber rigidity during spinning are thus eased by maintaining a reasonably warm and humid atmosphere.

The elastic properties discussed so far relate to stresses applied at relatively low rates. When forces are applied at rapid rates, then dynamic moduli are obtained. The energy relationships and the orders of magnitude of the data are much different [570]. Because of the experimental difficulties, only little work at rapid rates has been carried out with cotton fiber compared to that done with testing at low rates of application of stress. In contrast, cotton also responds to zero rate of loading, i.e., the application of a constant stress. Under this condition the fiber exhibits creep that is measured by determining fiber elongation at various intervals of time after the load has been applied. Creep is time-dependent and may be reversible upon removal of the load. However, even a low load applied to a fiber for a long period of time will cause the fiber to break.

7.4   ELECTRICAL PROPERTIES

Electrical properties of fibers were first considered important because of the effects associated with the build up of static charges that could hinder mechanical processing and with certain discomfort and hazards associated with electrical charges in clothing, carpets, upholstery, etc. The degree to which a material is susceptible to electrical charging is referred to as its electrical permittivity or dielectric constant. Another factor, closely related to dielectric constant is the electrical resistivity, which describes the degree to which electrical charges can be conducted through a material to which an electrical potential or voltage differential is applied [581].

In practice, the dielectric constant, ε_r, is determined from the measurements of electrical capacitance as the ratio C_p/C_o, where C_p is the capacitance with the material between the plates of the condenser and C_o is the capacitance of the empty space. The capacitance and thus dielectric constant of cotton is dependent on three parameters including electrical frequency, moisture content, and temperature. The dielectric constant of all materials decreases with increasing frequency. As the moisture content increases, dielectric increases. Thus, in the case of cotton at 0% RH, the dielectric constant ranges from 3.2 at 1 kHz to 3.0 at 100 kHz. At 65% RH, cotton's dielectric constant decreases from 18 at 1 kHz to 6.0 at 100 kHz [588].

The electric resistance of a material is defined as the ratio of the voltage applied across a material to the current (measured in amperes) that flows as a result. The unit of resistance is Ω (V/A). It is more convenient to use specific resistance r, defined as the resistance between opposite faces of a 1-m cube. However, as was the case for fibers, it is more convenient to base the measurement on the linear density rather than cross-sectional area. This leads to the mass specific resistance, $R_S = rd$, where d is mass density. R_S is usually expressed in Ω g/cm^2. Under standard conditions, R_S for raw cotton is approximately 0.5×10^6 Ω g/cm^2. As raw cotton is washed and otherwise purified, its resistance increases at least 50-fold [589]. Moisture content has even a greater role in resistance with $R_S = 10^{11}$ Ω g/cm^2 at 10% RH decreasing a million-fold to $R_S = 10^5$ Ω g/cm^2 at 90% RH [590].

7.5   ADVANCED FIBER INFORMATION SYSTEM

A recent development that incorporates several fundamental measures into one system is the advanced fiber information system (AFIS) [591–593]. The AFIS measures several fiber properties that are key to predicting the ease of spinning and quality of the finished product. Included in the measurements are fiber neps (small tangles of fiber), dust, trash, fiber length including short fiber, and maturity. The measurements are unique in that, as will be explained below, the technology is such that it automatically counts and sizes individual fibers and particles (neps and dust). The principle of operation is that a fiber individualizer aeromechanically opens and separates the sample into single fibers that are injected into an airstream. Dust and trash particles are diverted to a filter while the airstream transports the fibers and neps past an electro-optical sensor that is calibrated to measure the specific size characteristics of the fibers and neps. With regard to neps, AFIS determines their average size and size distribution. The measurements of dust and trash include the distribution of the size of dust and trash particles along with their number per gram and average size. The AFIS length measurements include percent of short fiber content (<0.50 in./12.7 mm) by number and by weight; average length by number and by weight; coefficient of variation of fiber length by number and by weight; upper quartile length (i.e., the length for which 75% of fibers are shorter) by weight; and the 5.0% length (i.e., the length for which 95% of the fibers are shorter) by number. Fiber maturity measurements include fiber fineness (linear density measured in millitex), and the immature fiber content (percentage of immature fibers by number).

8 Cotton Fiber Classification and Characterization

CONTENTS

Cotton bale weights vary from about 375 to 515 lb (170 to 233 kg) depending on the country in which they are produced (see Table 1.1). A pound of cotton contains 100 million or more individual fibers and each individual fiber varies in properties along its length from one end to the other. Cotton classification is a practical, cost-efficient way of measuring the general quality and physical attributes of bales of cotton fiber that affect the quality of the finished product and manufacturing efficiency and allows a market value to be established for the cotton to facilitate utilization.

Many varieties of the four domesticated species of cotton are produced commercially in widely varying locations and growing conditions throughout the world by thousands of farmers on millions of acres of land. The varying locations and conditions result in many different grades of cotton that vary widely in quality and fiber properties because of different varieties, soils, weather (rainfall, temperature, etc.), farming cultural practices (irrigation, fertilizers, crop protection products, etc.), insect damage, length of growing season, exposure of open cotton before harvest, methods of harvesting and ginning, storage, and a host of other variables [3,6,10,42,594–599]. Variations can exist not only between farms but also on the same farm and within a single bale. These considerations do not apply to synthetic fibers.

The quality of cotton is important to cotton farmers because it is related to the price and yield of cotton produced and provides farmers information to evaluate production, harvesting, and ginning practices and to market their cotton effectively. Quality is even more important to the users of cotton because it is highly related to efficiency in manufacturing and to the quality and utilization of yarns and fabrics produced, as well as to effecting the dyeing and finishing processes and outcome. For example, the color grade of raw cotton can affect dyeability if the cotton is not scoured and bleached adequately [600]. It is, therefore, of commercial importance that producers and users know the quality of cotton.

Because of the size and method of packaging, it is not practical for a buyer to inspect actual bales of cotton at the time of purchase. In general practice, market transactions are made on the basis of representative samples drawn from each bale. A tag or coupon showing the bale number or other identification normally accompanies the sample. This sample furnishes the specimen for classification.

There are two methods for estimating fiber quality, expert appraisal by a trained classer, and instrumentation. Many different classification systems are used worldwide in the various countries where cotton is grown. For example, the United States [601], China [3,594,602], Uzbekistan [603], and Egypt [594] all have different systems. All systems measure cotton quality using similar parameters, even though the actual classification scheme and classification designations may be different. The determinations, either manually or by instrument, may be different for individual parameters also. The quality parameters are the following:

1. Amount of foreign matter
2. Preparation (destructive effects of ginning)
3. Color and luster (discoloration)
4. Fiber length and regularity (fccl and appearance)
5. Fiber fineness
6. Strength
7. Growth area, and how ginned (in some countries these are considered the most important parameters)

Saw-ginned cottons (Upland cottons) are classed on a different system than roller-ginned cottons (longer-staple fibers and the very short Asiatic "desi" cottons).

The U.S. HVI system for the classification of fiber quality and instrumentation for measuring other fiber properties are becoming more widely used all over the industrialized world, as well as in some developing countries for determining the proper laydown and mix for processing cotton in textile mills. Therefore, the U.S. system will be described in more detail than the other classing systems. Information also is given on how the various quality parameters can affect cotton fiber chemistry.

The present HVI system as developed in the United States has been used principally for marketing of the U.S. cotton to the U.S. textile industry. With the disappearance of the U.S. textile industry there has been an impetus toward the standardized testing of cotton on a global basis [604]. The concern is that HVI measurements of length, strength, fineness, color, and trash were probably sufficient for marketing Upland cotton to an industry concentrating on production of medium weights primarily with rotor spinning, but are not satisfactory for handling the global market that is still primarily ring spinning of fine counts. To adequately support the marketing of cotton produced from around the world (wide diversity of varieties, growth conditions, and harvesting and ginning protocols), it would seem that several other fiber properties would be necessary for satisfactory fiber characterization. These would probably include short fiber content, fineness and maturity, neps, and stickiness. The establishment and management of such a global system offer several challenges including

1. Developing a system that contains a minimum number of measurements that are truly independent and are of demonstrated value for predicting process and product quality for global textile manufacturing.
2. Maintaining all testing stations at comparable testing levels.
3. Establishing a worldwide database of fiber properties

The USDA Agricultural Marketing Service has established such a marketing system for the U.S. cotton and should be the model and leader in establishing such a global marketing system. One of the immediate challenges is that many cotton-testing systems around the world are unable to afford to maintain exact environmental conditions for testing (70°F and 65% RH). One of the proposals to handle this is to supply HVI systems with built-in sensors that monitor the moisture content of the sample under test and automatically correct the test results to what they would be when testing at standard conditions.

8.1 U.S. CLASSIFICATION SYSTEM

In the United States, cotton classification is the process of describing the quality of cotton in terms of color grade, leaf grade, length, strength, and micronaire reading according to the official cotton standards [601]. The USDA Cotton Program in 2004 operated 240 HVI instruments for classifying the U.S. cotton crop. These instruments measure micronaire, length, length uniformity, strength, reflectance (Rd), yellowness (+b), and trash percent area. Grade has been shifted into color grade and leaf grade. The HVI color Rd and color +b values are combined into an official HVI color grade via the official USDA color chart. Work is in progress to replace the classer's visual grade with an HVI determined leaf grade. In recent years, USDA has classed about 98% of the cotton produced on the U.S. farms.

8.1.1 SAMPLING

A classification sample normally consists of two parts of about 85 g each taken from opposite sides of a bale. To obtain the sample, cuts 15 cm wide are made in the side of the bale before or after they are wrapped, deep enough to provide a representative specimen from the bale.

8.1.2 GRADE

The official cotton standards of the United States for the grade of Upland cotton are also called the universal standards. Leading cotton associations in major cotton-consuming countries meet periodically to establish a continuing consensus of cotton classification. International conferences are held every 3 years in the United States to consider revisions and to ensure accurate reproduction of the standards. By this method, the U.S. cotton classification system maintains sensitivity and responsiveness to cotton consumer needs.

Grade represents an assessment of two factors, color grade and leaf grade, which are now assessed and reported separately, the former derived from the HVI values for color Rd and color +b. Color grade and leaf grade standards, prepared and maintained by Cotton Division, Agricultural Marketing Service (AMS), provide a basis for maintaining consistency in the assessment of these grades. Color grade and leaf grade designations are shown in Table 8.1 through Table 8.3.

The Pima and the Upland grade standards differ. The Pima cotton is naturally of a deeper yellow color than the Upland cotton. The leaf contents of Pima standards are peculiar to this cotton and do not match that of Upland standards. Pima cotton is ginned on roller gins and has a stringy and lumpy appearance. Pima cotton grades range from the highest grade of 1 to the lowest grade of 10.

8.1.2.1 Color

The Upland cotton is naturally white. Continued exposure in the field to weathering and the action of microorganisms can cause the cotton to become darker. Under extreme conditions of weather damage, the color can become a very dark bluish gray. If growth is stopped prematurely by frost, drought, etc., the lint is characterized by increased yellowness. Cotton can also become discolored or spotted by the action of insects, microorganisms, and soil stains. Any departure from the bright color of normally opened cotton is interpreted to indicate deterioration in quality. In the classification of Upland cotton, these color differences are recognized, divided into categories, and described in terms of color grades (see Table 8.1).

8.1.2.2 Leaf and Other Trash

Cotton usually becomes contaminated by leaf and other trash in various amounts through exposure in the field and in harvesting. The amount of foreign matter remaining in the lint after ginning is largely dependent upon the trash content, the condition of the cotton at the time of harvest, and the number of cleaning and drying machinery used by the gin. Much of the foreign matter is removed by the cleaning and drying processes during ginning, but it is impractical to remove all of it.

TABLE 8.1
Color Grades of the U.S. Upland Cotton[a]

Color Grades	White	Light Spotted	Spotted	Tinged	Yellow Stained
Good middling (GM)	11*	12	13	—	—
Strict middling (SM)	21*	22	23*	24	25
Middling (M)	31*	32	33*	34	35
Strict low middling (SLM)	41*	42	43*	44	—
Low middling (LM)	51*	52	53*	54	—
Strict good ordinary (SGO)	61*	62	63*	—	—
Good ordinary (GO)	71*	—	—	—	—
Below grade (BG)	81*	82	83	84	85

[a]The first number of the 2-digit code for color grade is for GM (1) through BG (8) and the second number is for white (1) through yellow-tinged (5); e.g., 11 is good middling white, 33 is middling spotted.
*Physical standards, all others are descriptive.
Source: From The Classification of Cotton, in *Agricultural Handbook Number 566*, Agricultural Marketing Service, U.S. Department of Agriculture, Washington, D.C., April 2001.

TABLE 8.2
Relationship of Trash Measurement to Classer's Leaf Grade for the U.S. Cotton[a]

Trash Measurement (4-Year Average) (% Area)	Classer's Leaf Grade
0.13	1
0.20	2
0.34	3
0.51	4
0.72	5
1.00	6
1.25	7
1.57	8

[a]The surface of the cotton sample is scanned by a video camera and the percentage of the surface area occupied by trash particles is calculated. The trash determination and classer's leaf grade are not the same but there is a correlation between the two as shown in this table.
Source: From The Classification of Cotton, in *Agricultural Handbook Number 566*, Agricultural Marketing Service, U.S. Department of Agriculture, Washington, D.C., April 2001.

TABLE 8.3
Percentage Distribution of Color Grade and Leaf Grade, the U.S. Cotton— 2003 Crop

	Leaf Grade					
Color Grade	1&2	3	4	5	6	7
11 & 21	12.5	9.6	0.8	*	*	*
31	3.3	29.8	11.1	0.6	*	*
41	0.3	13.0	11.0	1.1	0.1	*
51	*	0.4	0.6	0.1	*	*
61	*	*	*	*	*	*
71	*	*	*	*	*	*
12 & 22	0.3	0.5	0.41	0.06	*	*
32	0.1	0.8	2.17	0.68	*	*
42	*	0.6	4.23	1.14	*	*
52	*	0.1	1.01	0.28	*	*
62	*	*	*	*	*	*
13 & 23	*	0.03	*	*	*	*
33	*	0.18	0.1	*	*	*
43	*	0.35	0.1	*	*	*
53	*	0.20	*	*	*	*
63	*	0.02	*	*	*	*

*Less than 0.01% of crop.
Source: From Cotton Quality Crop of 2003. U.S. Department of Agriculture, Agricultural Marketing Service, Cotton Division, Memphis, TN, Vol. 77, No. 7, pp. 10–11, May 2004.

Leaf includes dried and broken plant foliage of various kinds and may be divided into two general groups:

1. Large leaf
2. "Pin" or pepper leaf

Large leaf is usually less objectionable because large particles are more easily removed by the textile manufacturing process. Leaf and other foreign matter in cotton must be removed as waste in the textile manufacturing process either in mechanical or in wet processing. Small particles of foreign matter that are not removed in manufacturing detract from the quality and appearance of the manufactured yarn and fabric. Cottons with the smallest amount of foreign matter, other properties kept equal, have the highest value. In the classification of the U.S. cotton, these differences are described in terms of leaf grade (see Table 8.2). Bark, parts of seeds, shale (lining of the bur), motes (immature undeveloped seeds), grass, sand, oil, and dust are also sometimes found in ginned cotton. Research is in progress to replace the classer's visually measured leaf grade with an instrument (HVI) measured leaf grade.

8.1.2.3 Preparation

Preparation describes the relative appearance of the ginned lint in terms of smoothness or roughness. Various methods of harvesting, handling, and ginning cotton produce differences in roughness or smoothness of preparation that sometimes are very apparent. Laboratory tests do not show that these differences in preparation in the raw cotton result in important differences in spinning quality; however, generally, cotton with a smooth appearance introduces less waste than cotton with a rough appearance. Longer cottons normally will have a rougher appearance after ginning than shorter cottons, but that does not necessarily mean that yarns made from such cottons will be relatively poorer.

8.1.2.4 Neps

Neps are small, entangled clumps of fibers caused by mechanical processing. They are visible as dots or specks when a thin web of fibers is held to the light or against a dark background. Neps in lint are undesirable because they will appear as defects in yarns and fabrics. The removal of neps from the lint is difficult, costly, and sometimes impossible. The longer, finer cottons tend to have more neps than the shorter, coarser cottons. Lint having a high percentage of thin-walled, immature fibers is especially likely to be neppy, which can lead to whitish flicks or specks on the fabric after dyeing. Neps are difficult for the classer to detect or evaluate in classification.

8.1.2.5 Special Conditions

The USDA cotton classing regulations provide for special designations because of the presence of extraneous matter or other irregularities that cause the usefulness of the cotton to be below that normally expected. When such a designation is made, it is reported as a comment in addition to the normal grade information. Notations are normally made on classification memoranda and provide a basis for segregating bales with special conditions from normal bales. Conditions that require special comments include rough preparation, gin-cut or reginned cotton, repacked, false packed, mixed packed, or water-packed cotton, and excessive grass, bark, or extraneous matter.

8.1.3 LENGTH

Fiber length is determined by instrument measurement of a tuft of fibers prepared automatically by the mechanical sampler. The length distribution of the fibers in the specimen is determined by scanning the tuft of fibers to produce a fibrogram. Statistical properties of the fibrogram are then used to determine two length parameters for use in assigning length values to the bale from which the specimen was selected. The upper half mean length (UHM) is the average length of the longest half by number of the fibers in the specimen. The mean length (ML) is the mean length of all the fibers. Values actually reported for classification are UHM and length uniformity index (UI), which is the percentage ratio of ML and UHM. These two values, UHM and UI (measured by closely calibrated instruments), are used for length classification and reported as staple length.

8.1.4 STRENGTH

Fiber strength measurements are made on the same specimen used in making the length measurement. After the length measurement is completed, the specimen is repositioned to a point of constant mass as determined by the instrument, clamped in two sets of jaws, and broken by transverse motion of one of the sets of jaws. A sensitive transducer attached to the jaws provides a very accurate measurement of the force required to break the specimen. Since, theoretically, the mass of the specimen at the breaking point is held constant, the grams force required to break the specimen is directly related to the tenacity of the specimen. The tenacity measurement is expressed as grams force (gf) per unit mass (tex), where tex is the weight in grams of a specimen having 1000 m of length. Tenacity of the Upland cottons ranges from approximately 20 to 30 gf/tex.

8.1.5 LINEAR DENSITY

The linear density of the fibers is an important factor in determining the processing performance, notably the spinning performance, and the quality of cotton yarns. Linear density is typically reported in millitex (mass in milligrams of 1000 m of fiber). For a complete definition of linear density and fiber fineness characteristics, measurement of both fineness and maturity are required. Although methods for measuring these two factors separately are studied, no practical method has been accepted for use in cotton classification. Fineness and maturity in combination are measured by resistance to airflow. A specimen of specific weight is compressed to a specific volume in a porous chamber. Air is forced through the specimen and the resistance to airflow is related to the specific surface area of the fibers and is a function of both the fiber linear density (fineness) and maturity. The property so measured, called micronaire, has been used for a number of years and is an integral part of the U.S. classification system, and provides a measure of, but is not equivalent to, fiber linear density. The micronaire reading is simple, low cost, and fast and is one of the most useful quality measurements in cotton classification. Typical micronaire readings for good-quality Upland Cottons range from 3.5 to 5.0 but vary from year to year due to growing conditions, the most desirable range between about 3.5 and 4.1. When instruments are developed that are capable of measuring fineness and maturity independently and rapidly, they will likely be adopted into the classification system quickly.

8.1.6 COLOR (MEASUREMENT OF REFLECTANCE AND YELLOWNESS)

The Upland cotton is naturally white in color when it opens under normal growing conditions in the field. A number of factors influence changes in its color after the cotton has opened.

Whatever the source of the change, color reduction from the characteristic white observed for newly opened cotton is likely to be an indicator of deterioration in quality. Color differences in cotton can be a source of variations in dye shades in finished fabric. Color in raw cotton is measured by the cotton colorimeter. Two characteristics are measured—the reflectance (Rd) and the yellowness (+b)—under standard illumination and density. In the U.S. system, these two values are used to determine the color grade, which was previously determined visually.

8.1.7 TRASH CONTENT (MEASUREMENT)

Trash content, or foreign matter, includes all materials in the cotton sample except the lint. Types of trash possibly present in cotton have been listed previously. Instrument systems used for cotton classification include image analysis components consisting of high-resolution video cameras and computer hardware and software for interpreting video image scans of cotton samples. The dark contrast of the trash particles against the lighter background of the cotton lint allows these instruments to count the number of particles, measure their sizes, and determine what percentage of the specimen surface area is covered by the trash particles. This percentage area is related to the gravimetric weight of trash in the cotton sample. Percent area measurements of trash content range from 0.1 for very clean cotton to 0.9 for cotton with very high trash content (see Table 8.2). These measurements are reported in addition to the leaf grade assigned from visual observations, work in progress to replace the classer's leaf grade with an instrument (HVI) leaf grade.

8.1.8 REPORTING COTTON CLASSIFICATION RESULTS

Cotton classification services are provided to the U.S. cotton growers by Cotton Division, AMS, USDA on a fee basis. The classification data belongs to the cotton growers and is reported to them by various mechanisms (including a card showing all test values or electronic data transfer and printout). The classification data is maintained in a national data bank by the AMS. As the cotton moves through marketing channels, classification data is made available to the owner of the cotton by electronic data transfer. To obtain the data, the proof of ownership and a small fee are required. As measurements become more precise and accurate and industry develops confidence in the test results, it is anticipated that all cotton users will rely on test information from the national data bank for marketing and utilization of the U.S. cotton.

8.2 CLASSIFICATION SYSTEMS OF SOME COUNTRIES OTHER THAN THE UNITED STATES

There is a worldwide movement toward the increased use of instrument measurements such as the HVI in characterizing cotton quality and in cotton classification. This is encouraged by international bodies such as the International Cotton Advisory Committee (ICAC) and the International Textile Manufacturers Federation (ITMF).

8.2.1 CHINA

In China [3,10,594,602], cotton is grown on small plots, handpicked, and harvested in sacks. The seed cotton is weighed, sampled, and graded by the government classer (first classing). Moisture, staple length, and lint turnout are determined, and similar lots—by one of the seven grades—are combined and ginned according to class. The saw-ginned cotton is classed again with a system very similar to the U.S. manual system previously used. The Chinese use a three-digit numbering system to classify lint. The first digit is for the grade, which is determined by color, maturity, and ginning quality of the cotton. The second and the third digits are the staple length in millimeter. For example, Type 327 would be grade 3 (under the U.S. system this would be middling color, about leaf grade 4 (see Table 8.1 and Table 8.3), 27 mm (equivalent to 1 1/32 in.). In addition, moisture content, trash content, and bale weight are also determined in classification. In 2003, China unveiled a cotton classification reform plan aimed at transitioning from the current manual classification system to an instrument-based classification system within 5 years. The reform plan includes transitioning to UD bales, two-sided bale sampling, and HVI inspection of all Chinese cotton (bale form). The China Fiber Inspection Bureau, responsible for cotton classification, has acquired 75 HVI systems in 2004 and 2005 as part of a process to shift from a hand-class system, based on grades and staple length to the U.S. style HVI system. A color chart for grading Chinese cotton and a method to determine the gravimetric trash content are under development.

8.2.2 CENTRAL ASIA REPUBLICS

In the former Soviet Union [603], there were two classification systems. One was applied when supplying cotton to the domestic market and other socialist countries, the other for cotton sold to hard currency markets. The domestic system was GOST 3279-76. It had no scientific basis and could not estimate satisfactorily fiber characteristics for yarn production. The export version had no officially recognized physical standards. So a new system somewhat approaching the American system that could be used in international markets was developed.

In 1993, a new classification system for cotton fiber quality, RST Uz 604-93, was introduced in Uzbekistan [603]. Since 1995, this standard has also been used in Kazakhstan and Tadjikistan [603]. In Azerbaijan, attempts are made to classify its cotton in accordance with international standards [605].

The Uzbekistan classification of cotton fiber is based on three quality levels: types, sorts, and grades, which are determined with instruments (including HVI type) and classer's methods, they have some 35 HVI systems in place at gins and at Tashkent, all operated by GOST. Their cotton is still traded, however, on the traditional hand-class standards basis established by GOST. Types of cotton are based on length, linear density, and strength (g/tex). The basis of division into types is fiber length in millimeter. There are nine types: 1a to 3 are long staple cottons; 4 to 7 are medium stable. Cotton sort is defined on color and maturity. There are five sorts bearing the names of ordinal numbers in Uzbeki: Birinchi, Ikinchi, Uchinchi, Turchinchi, and Beshinchi. Birinchi sort represents medium staple cotton with white color, no spots, and maturity greater than or equal to 1.8 (CIS method; about 3.6 micronaire, maturity index 0.85). Ikinchi sort has pale yellow spots (like very light spotted) and lower maturity (up to 1.6). In Uchinchi sort, the yellow spots (similar to spotted) are more intense and maturity lower to 1.4. Turchinchi is creamy with brown stains (like tinged) and the maturity lower to 1.2. Beshinchi is mat-grey with brown stains and maturity is very low. Cotton grade is based on preparation and trash content. Within Birinchi and Ikinchi sorts, there are five grades, within Uchinchi and Turchinchi four grades, and within Beshinchi three grades. Altogether, there are 21 grades, which are called Oliy, Yakshi, Urta, Oddiy, and Iflos in Uzbeki (i.e., high, good, middle, ordinary, and contaminated, respectively). The new cotton fiber classification in Uzbekistan is comparable to USDA classification.

Uzbekistan classification and differentiation of cotton is effected immediately after the harvest. At present, cotton quality evaluation is carried out by cotton mill laboratories. This is felt to be undesirable (because of the effect on competitiveness and prices). Cotton is classified as seed cotton before ginning and as fiber after ginning (as described above). The picked raw cotton is taken to collecting points where the classers and laboratories carry out evaluation of raw cotton fiber quality. They sort fiber according to uniform fiber quality features: origin (according to farms), type and sort, trash content, and moisture content. To regulate receiving and classing of raw cotton in points of preliminary preparation, there is a special government standard for raw cotton (RST Uz 615-94 standard: Raw Cotton, Technical Conditions, introduced in 1994). Grade 1 is manually picked cotton; grade 2 is mechanically harvested cotton and also manually picked but contains more trash than grade 1; grade 3 is cotton picked from the ground and contaminated cotton of manual and mechanical picking, which contains more trash than grade 2. The seed cotton classing is meant to stimulate producers to adopt practices to lower the contamination of raw cotton. Also ginning of large quantities of picked cotton of uniform quality produces uniform fiber, which in some cases is sold without fiber quality evaluation.

8.2.3   FRANCOPHONE AFRICA

For most of the former French colonies that are now independent countries in Africa, as well as in many other African countries, cotton cultivation management, ginning, and fiber marketing are operated by one company [606]. Cotton classing is also operated by the company's classing office, and is based on types and grades. Some HVI lines are in operation but are not used for classing *per se*. Samples of each bale sorted by gin plant are conditioned over several hours. Classing is performed by trained classers visually and micronaire measurements are made by instrument. Length is estimated with the pulling method and expressed in 1/32 in. Grade is visually estimated according to the color, trash, and preparation. Color and trash are not considered separately. There is a specific grade standard system for cottons of French-speaking countries in Africa. The system is based on six grades 0, 1, 2, 3, 4, 5, from best to worst. These standards are the base for determination of sales types specific to each country. Bales' marking is done at the gin according to a codification for length and another for grade.

8.2.4   SUDAN

In Sudan [607], cotton is classified twice, as seed cotton before the gin and as lint at the port of shipment. The handpicked cotton is piled up on covers where trash and other contaminants are removed. Grades are assigned to the piles and the graded cotton is transported to the gin. The field classification system is felt to be key to implementing quality control measures such as timely and early picking, which can reduce stickiness. The ginned cotton is baled and transported to Port Sudan where the lint is reclassified and samples are taken for fiber testing.

8.2.5   SOUTH AFRICA

In South Africa [608] prior to ginning, producers submit samples of seed cotton both handpicked and machine picked for classification processes by using RSA Seed Cotton Standards for grade to obtain information on the expected quality of their crop. When buying seed cotton, it also determines the price paid to the farmer whether buying or contract ginning; accurate preselection of seed cotton and homogenous blending of lots prior to ginning is viewed by ginners as important to ensure that the lots of cotton lint are uniform in grade and quality. Ginners are currently extensively using HVI evaluation, and the seed cotton grading function is viewed by ginners as a control measure.

The Quality Control Division of Cotton, South Africa, is mainly responsible for the grading and classification of the South African cotton crop but this is not a compulsory service. Ginners who participate submit samples from each bale to receive a grading certificate that is used for the local marketing of their crop. Ginners who are responsible for their own grading and classification are required to use the same grade and quality specifications.

The Quality Control Division uses HVI for fiber analysis. The system measures color, trash, fineness, length, strength, elongation, and length uniformity of cotton. The determination of grade is done visually in accordance with the South African lint standards, namely Deal, Dirk, Doly, Duns, and Lfy, which are comparable to U.S. lint standards good middling, strict middling, middling, strict low middling, and low middling, respectively. (As the U.S. system now separates grade into color grade and leaf grade, these grades would be equivalent to the U.S. color grade with about a leaf grade 3 (See Table 8.1 and Table 8.3.)) The South African lint standards for grade with regard to color and trash are similar to the (white) Universal Standards for Grade of the American Upland Cotton, except that provision is made for soil stained and yellow spot in the lower South African standards (Duns and Lfy). As a certain percentage of the South African crop is handpicked, South African cotton tends to be a brighter white than the U.S. standards.

8.2.6 EGYPT

The Egyptian cotton [594] is classed by the appraisal of experts based on their own standards. The standards are meant to illustrate color, lightness, immature and dead fiber content, foreign matter content, and preparation of the cotton. There is no separate classing of length: length is deduced from the variety. An exact comparison of the Egyptian and the American systems of classing is impossible because of their differing principles. However, when only foreign matter content, preparation, and to some extent color are taken into account, the grades can be related.

8.2.7 INDIA

The Indian cotton [3,594] is classed in a way resembling the Egyptian system and not based on the American system. The system takes into account species, variety, grade, and length. Place of cultivation and ginning method are also mentioned. East India Cotton Association prepares and maintains grades and staple standards. There are five principal grades: extra superfine, superfine, fine (basis), fully good, and good with staple lengths varying from 17 to 40 mm 22/32 to 1 18/32 in.) and are roughly comparable to the U.S. standards of strict low middling to good ordinary (see Table 8.1 and Table 8.3).

8.2.8 PAKISTAN

The situation in Pakistan [3,609] is similar to that in India; in Pakistan, trading is based upon hand-classed variety and grade. Little attention has been paid to the cleanliness of the product. The crop is not graded, and cleaning at the gin (there are few seed cotton and lint cleaners) has been largely neglected [590]. However, this is changing. HVI systems have been introduced for conducting tests and preparation of standards for cotton grades [609].

8.2.9 AUSTRALIA

The cotton classing in Australia [610] encompasses elements of the old and revised USDA system. Qualified cotton classers classify the cotton by examining three characteristics: grade, staple, and micronaire, and HVI is widely used as a guide by classers. Micronaire is determined with an airflow instrument. There are 44 different grades of cotton based on color of the fiber, trash content, preparation, and staple length (which has 3 different standards). Length is reported as

1/32 of an in. with color and leaf grades determined by means of USDA standard grade boxes, and color and leaf grades are reported separately. Other cotton fiber properties (strength, length uniformity, elongation, stickiness, nep count, and moisture content) are tested to determine the fiber manufacturing potential, many of which are measured on HVI systems.

8.3 IMPACT AND FUTURE OF HVI SYSTEMS

Hunter [611] has presented an excellent review of literature relevant to theoretical or empirical relationships and fiber quality indices (FQIs) that have been developed to predict yarn quality, especially strength and evenness, from the measurement of raw fiber properties. Theoretical approaches are based upon fundamental principles of physics and mechanics but due to the complexity of the mechanisms of yarn formation, doubt is generally raised as to the practicality of such approaches to reliable predictions of textile quality. Empirical relationships are primarily based on linear- or multiple-regression analysis approaches, which work well for the particular cottons in the data set but have limited usefulness when attempting to expand beyond the particular range of cottons covered. More recently, many approaches including artificial neural networks and fuzzy logic have been used for modeling of textile quality from fiber properties . However, Hunter concludes that, as of yet, there is no viable solution. Hunter's review shares an interesting breakdown of the relative contributions of HVI fiber properties to predicting the strengths of rotor versus ring yarns. From Table 8.4 we see that strength, elongation, and color play a more significant role in rotor yarn strength, whereas length and length uniformity play a more significant role in ring yarn strength.

Suh and Sasser [612] have reviewed the impact of HVI systems on both the domestic and world cotton textile industries. HVI has totally revolutionized the way in which cotton is marketed and processed in the modern textile industry. By purchasing cotton bales based on HVI measurements the textile industry is now able to systematically store, retrieve, and form multibale input laydowns designed for uniformity of process and product quality. In addition to product uniformity, HVI data allows for selection of appropriate raw materials at reduced costs while maximizing product quality and profit. Despite the obvious assets of HVI, at this time, it still does not produce information of certain fiber properties that are critical to process optimization and control. Included among these are true short fiber content, fineness and maturity, seed coat fragments, and sugar. It is hoped that research now carried out, especially in the United States, will lead to further enhancement of HVI systems.

TABLE 8.4
Relative Contributions of HVI Fiber Properties in Predicting the Strengths of Rotor versus Ring Yarns

HVI Property	% of Property Contributing to Rotor Yarn Strength	% of Property Contributing to Ring Yarn Strength
Strength	24	20
Length uniformity	17	20
Length	12	22
Micronaire	14	15
Elongation	8	5
Color or Reflectance	6	3
Unexplained	13	12

In December 2003, the ICAC formed an Expert Panel on Commercial Standardization of Instrument Testing of Cotton (CSITC). The purpose of the CSITC is to promote instrument testing of cotton on a global basis. The driving force behind this is to enhance the competitiveness of cotton with synthetic fibers in the global market place. At a meeting at Mumbai, India, in 2004, the panel set for itself several goals to encourage worldwide testing of cotton with standardized instruments. These included:

1. Definition of specifications for cotton trading
2. Definition of international test rules
3. Implementation of test rules
4. Certification of testing laboratories
5. Definition and provision of calibration standards
6. Specification of commercial control limits for trading
7. Establishment of arbitration procedures

The two primary agencies proposed to help implement this program are the Bremen Fiber Institute and the USDA's AMS Cotton Program [613].

9 Production, Consumption, Markets, and Applications

CONTENTS

9.1 PRODUCTION, CONSUMPTION, AND MARKETS

Cotton is the most important natural vegetable textile fiber used in the world for spinning to produce apparel, home furnishings, and industrial products [614,615]. It continues to be one of the most dominant textile fibers in much of the world and accounted for over 38% of the worldwide textile fiber consumption in 2005 [1] (see Table 9.1). China, India, Pakistan, Turkey, the United States, and Brazil are the major cotton-consuming countries [40,614] (see Table 9.2). Consumption is measured by the amount of raw cotton fiber, the textile mills purchase and use to manufacture textile materials. China, the United States, India, Pakistan, Uzbekistan, Turkey, and Brazil are the major cotton producing countries, accounting for over 81% of world cotton production in 2004 [40,614] (Table 9.3). Organic cotton, which is a very small niche market, was produced in at least 16 countries in 2003, with Turkey as the leading producer (Table 9.4).

In addition to the various markets for cotton lint, there are also markets for cottonseed and its products [616]. Cottonseed represents about 15–20% of the total value of cotton. Vegetable oil for human consumption, whole cottonseed, meal, hulls for animal feed, and linters for batting and chemical cellulose are the major end uses for cottonseed [616].

Since 1988, the trend has returned to natural fibers in the United States and Europe. National surveys of U.S. consumer attitudes about cotton versus synthetic or manufactured fibers show that consumers think cotton has substantial advantages over polyester regarding functional and quality attributes. The difference is greatest regarding comfort, but substantially more consumers also believe that cotton offers better overall quality. This has created an increased demand for fabrics and garments of 100% cotton and blends with high cotton content. Previously, most blended fabric contained about 65% polyester; now the trend is toward reverse blends (60% cotton and 40% polyester or 85% cotton and 15% polyester) with higher cotton content. However, in the less developed countries polyester usage is increasing faster than cotton.

The textile industry in most European countries and the United States has become smaller or almost totally disappeared. China, India, Pakistan, and Turkey are the major textile producers. In both the United States and Europe, textile imports have greatly increased. The textile and apparel industries appear to be moving from the highly industrial countries to

TABLE 9.1
World Production of Textile Fibers (Million Lbs.)

Fiber	2000	2001	2002	2003	2004
Cotton[a]	42,628.3	47,404.8	42,376.8	45,653.3	57,528.5
Man-made fibers	62,685.3	62,446.4	66,526.7	69,893.1	75,409.9
Synthetics	57,801.7	57,855.3	61,842.8	64,886.0	69,916.7
multifilament and monofilament yarns	32,564.1	32,997.3	35,257.7	37,219.1	40,259.1
staple, tow, & fiberfill[b]	25,237.6	24,858.0	26,585.0	27,666.9	29,657.5
acrylic	5,806.9	5,647.5	5,981.9	5,937.0	6,047.8
multifilament & monofilament yarns	11.0	11.0	11.0	11.0	5.9
staple, tow, and fiberfill	5,795.9	5,636.5	5,970.9	5,926.0	6,041.9
nylon	9,075.6	8,341.5	8,688.3	8,728.0	8,996.5
multifilament & monofilament yarns	7,944.9	7,404.1	7,694.7	7,747.8	8,034.0
staple, tow, & fiberfill	1,130.7	937.4	993.6	980.2	962.5
polyester	42,228.5	43,127.5	46,402.0	49,289.7	53,804.6
multifilament & monofilament yarns	24,148.1	25,077.3	27,024.2	28,835.9	31,489.0
staple, tow, and fiberfill	18,080.4	18,050.2	19,377.8	20,453.8	22,315.6
other (except olefin)	690.7	738.8	770.5	931.3	1,067.7
multifilament & monofilament yarns	460.1	504.9	527.8	624.4	730.2
staple, tow, and fiberfill	230.6	233.9	242.7	306.9	337.5
Cellulosics	4,883.6	4,591.1	4,683.9	5,007.1	5,493.2
multifilament and monofilament yarns	1,109.1	1,089.3	1,022.7	1,080.7	1,063.3
staple, tow, & fiberfill[b]	3,774.5	3,501.8	3,661.2	3,962.4	4,429.9
Wool (scoured or cleaned basis)	3,042.0	3,000.0	2,884.0	2,659.0	2,687.0
Silk	N/A	N/A	N/A	N/A	N/A
Total	108,355.6	112,851.2	111,787.5	118,205.4	135,625.4
% Cotton	39.3%	42.0%	37.9%	38.6%	42.4%

[a]2003 Forecast.
[b]Excludes acetate cigarette filter tow.
Source: *Fiber Organon*, 76(7), July 2005 (Fiber Economics Bureau, Inc., Arlington, VA, USA)
N/A-no data available

the developing countries. The high imports do show that demand for cotton textiles in the United States and Europe continues to be strong at the consumer level.

9.2   APPLICATIONS

Cotton fabrics combine durability with attractive wearing qualities and comfort [617]. Cotton is inherently strong because the convolutions create friction within the fabric that prevents fibers from slipping. The wet cotton fabric is stronger than the dry cotton fabric. Cotton can withstand repeated washings and is ideal for household goods and garments that must be laundered often. A cotton garment on laundering may shrink because of the tensions introduced by spinning and weaving, but the cotton fiber itself is dimensionally stable and does not contribute significantly to the shrinkage. Cotton fabric will wrinkle and crease. However, cotton fabric can be treated to impart wrinkle resistance and dimensional stability as was discussed earlier in the book. Cotton can be dyed easily with a wide range of colors [618]. Cotton cellulose is not affected unduly by moderate heat, so that cotton fabrics can be ironed with a hot iron without damage.

TABLE 9.2
World Cotton Consumption (1000 U.S. 480 lb bales)

Country	1990/91	1995/96	2000/01	2004/05	2005/06
China	19,987	19,388	23,485	38,476	45,471
India	8,956	11,969	13,535	14,791	16,490
Pakistan	5,644	7,195	8,095	10,743	11,743
Turkey	2,478	4,360	5,164	7,095	6,896
United States	8,652	10,640	8,856	6,689	5,996
Brazil	3,319	3,757	4,197	4,197	3,997
Indonesia	1,492	2,135	2,448	2,249	2,299
Thailand	1,505	1,424	1,649	2,149	2,124
Bangladesh	446	551	999	1,874	2,074
Mexico	767	1,099	2,099	2,099	1,999
Russia	5,466	1,149	1,599	1,424	1,499
Korea, South	2,000	1,666	1,449	1,324	1,199
Taiwan	1,588	1,397	1,199	1,199	1,199
Egypt	1,456	1,009	750	949	999
Uzbekistan	839	872	1,024	874	799
Italy	1,469	1,538	1,329	924	750
Vietnam	188	170	430	675	750
Japan	3,025	1,528	1,189	814	725
Syria	250	375	550	700	725
Argentina	615	459	350	620	620
Other	15,328	13,054	11,704	8,963	8,504
Total	**85,470**	**85,736**	**92,100**	**108,828**	**116,855**

Source: USDA, Foreign Agricultural Service, April 2006. Crop year runs from August 1 to July 31.

Cotton fabrics are cool in hot weather. Although cotton is used as a fabric for hot-weather wear, it is also able to provide warmth. The warmth of the garment depends very largely on the pockets of air that are entrapped between the fibers in the fabric and the resistance of the fabric to the wind. Cotton fibers make good insulators when made into padded or quilted garments. Much of the comfort of a textile material depends upon its ability to absorb and desorb moisture. Moisture from perspiration collects in and passes through clothing as worn, and the properties of fabrics influence both the collection and passage of this moisture. Dynamic surface wetness of fabrics correlate with skin contact comfort in wear for a variety of fabric types, suggesting that mobility of thin films of condensed moisture is an important element of wearing comfort. Dynamic surface wetness of fabrics can be used to show why the cotton clothing is considered more comfortable than the clothing made of synthetic fibers [619]. A garment that does not absorb moisture, like garments made from synthetic fibers, will tend to feel clammy as perspiration condenses on it from the skin. Cotton fibers are able to absorb appreciable amounts of moisture, and having done so will get rid of it readily to the air. Cotton garments are, therefore, comfortable and cool, passing on the perspiration from the body into the surrounding air. No matter how tightly woven a cotton fabric may be, it will permit the body to breathe in this way. In addition, the absorbency of cotton makes it an excellent material for household fabrics such as sheets and towels. Static electricity is not a problem with cotton so the clothes do not cling to the wearer or to each other.

TABLE 9.3
World Cotton Production (1000 U.S. 480 lb bales)

Country	1990/91	1995/96	2000/01	2004/05	2005/06
China	20,687	21,886	20,287	28,982	26,183
United States	15,495	17,889	17,177	23,236	23,885
India	9,129	13,242	10,924	18,988	18,288
Pakistan	7,517	8,195	8,195	11,293	9,744
Uzbekistan	7,312	5,736	4,397	5,197	5,596
Brazil	3,291	1,883	4,309	5,896	4,697
Turkey	3,005	3,909	3,598	4,147	3,548
Australia	1,988	1,969	3,698	2,998	2,598
Greece	965	2,066	2,034	1,799	1,974
Syria	666	1,009	1,674	1,599	1,549
Burkina	354	294	525	1,179	1,324
Mali	527	775	480	1,099	1,149
Turkmenistan	2,006	1,149	824	919	974
Egypt	1,377	1,087	919	1,331	949
Kazakhstan	468	340	400	680	675
Mexico	856	973	394	625	635
Tajikistan	1,175	550	485	799	625
Tanzania	220	377	188	525	575
Argentina	1,354	2,089	758	675	550
Iran	547	799	735	615	550
Other	8,147	7,448	6,795	7,839	7,442
Total	**87,086**	**93,663**	**88,794**	**120,420**	**113,510**

Source: USDA, Foreign Agricultural Service, April 2006. Crop year runs from August 1 to July 31.

Cotton can be used in making rainwear fabrics. It can be woven tightly to keep out the driving wind and rain, yet the fabric will allow perspiration to escape. Special rainwear materials are woven in such a way that water swells the cotton fibers and closes up the interstices in the cloth.

The versatility of cotton has made it into one of the most valuable and most widely used of all textile fibers. Wherever a fabric is needed that is strong, hardwearing, and versatile, cotton can be used. There are literally thousands of actual uses (about 100 major uses) for cotton in textile items, ranging from baby diapers to the most fashionable dresses, coats, and jackets [617]. These uses can be classified into three main categories: apparel, home furnishings, and industrial.

In the apparel market, men's and boys' trousers and shorts use the largest amount of cotton, followed respectively by shirts worn by men and boys and men's and boys' underwear. Cotton is the major fiber for jeans, men's and boys' underwear and shirts. Denim fabrics utilize more cotton than any other single apparel item for trousers worn by men and boys. [617].

In the home furnishings, towels and washcloths account for the largest amount of cotton used, and sheets and pillowcases account for the second largest amount. The dominant fiber in towels and washcloths is cotton, supplying better than 90% of the market. It also holds almost half of the market for sheets and pillowcases [617].

Industrial products containing cotton are as diverse as tarpaulins, bookbindings, and zipper tapes, and cotton products can be used for clean up of agrochemical spills [620] and oil spills [621]. Some of the major industrial markets for cotton are medical supplies and industrial thread.

TABLE 9.4
World Organic Cotton Production [Metric Tons]

Country	1990/91	1991/92	1992/93	1993/94	1994/95	1995/96	1996/97	1997/98	1998/99	1999/00	2000/01	2001/02	2002/03	2003/04	2004/05
Argentina					75	75	300	300							
Australia			500	500	750	400									
Benin		45					1	5	20	20	30	38	46	25	67
Brazil				1	5	1	1	1	5	10	20				
Burkina Faso															45
China (PRC)	14											106	596	1,601	1,870
Egypt			50	153	600	650	625	500	360	200	200	200	122	122	240
Greece					300	150	125	100	75	50	50				
India			200	250	400	925	850	1,000	825	1,150	1,000	696	855	2,231	6,320
Israel												<425	390	380	436
Kenya								5	5	5					
Kyrgyzstan															2
Mali													19	35	65
Mozambique						100	75	50							
Nicaragua					20	20	20	20							
Pakistan				100	75	50	50	50					256	400	600
Paraguay													9	60	70
Peru			200	675	900	900	900	650	650	500	550	300	300	404	813
Senegal						1	1	10	50	146	200	-	6	6	27
Tanzania						30	30	100	230	190	180	400	380	600	1,213
Turkey			789	200	463	725	933	1,000	835	7,840	7,697	5,504	12,865	11,625	10,460
Uganda					25	75	75	450	250	246	248	250	500	740	900
USA	330	820	2,155	4,274	5,365	7,425	3,396	2,852	1,878	2,955	1,860	2,227	1,571	1,041	1,968
Zambia												2	3		
Zimbabwe								1	5	5					2
Total	**344**	**865**	**3,894**	**6,153**	**8,978**	**11,527**	**7,382**	**7,094**	**5,188**	**13,317**	**12,035**	**10,148**	**19,270**	**17,645**	**25,394**

Source: Organic Exchange, 2005–06 and ICAC, 2005–06 for most of the data; and Baird Garrott, Paul Reinhart AG, for Turkey, Israel, Mali and Burkina Faso in 2004/05. Crop year August 1 to July 31.

As a raw material for military items, cotton equips and helps support the military of every country. Many textile items are required by the armed forces in war and peace. It has been recognized by leading authorities on textiles and the national defense that too great a dependence upon fibers drawn from petrochemical feedstocks could present undesirable hazards to the military services from a supply standpoint [622,623]. A strong, viable cotton industry is essential to the defense capabilities of every country.

9.3 FUTURE TRENDS

Future demand in the world should remain high for cotton. The near term estimates of production and consumption are good [614,624] (Table 9.2 and Table 9.3). There is increased production in India, Pakistan, Turkey, and Brazil, and the world consumption in 2005 is expected to exceed production [614,625].

Much research on cotton is carried out in the United States and the rest of the world. The quality improvements in cotton continue to be significant. In the United States, for example, cotton breeders, farmers, and processors have responded to the demands of new technology by developing, producing, and delivering the longest, whitest, finest, and strongest cotton fiber ever grown. These trends should continue with new cottons, developed through biotechnology for insect resistance, herbicide tolerance, and improved fiber properties (e.g., stronger, finer, whiter fibers) as well as hybrid cottons and better new cotton varieties by conventional breeding (see Chapter 1). These new cottons are expected to have better fiber properties and require less crop protection products. Also, shorter-season, high-yielding plants are expected in the future. With these cottons, better yarns and fabric will be able to be produced. Cotton finishing will continue to be improved and open up new markets for cotton (e.g., rugs, carpets, nonwovens [626]). With respect to fiber quality, improved color fastness and evenness or levelness of dyeing, durability, and easy care characteristics look promising. Higher strength cotton can open new opportunities in industrial textile applications. New technology virtually has the potential to eliminate fiber contaminants before they reach customers. All of this research should help the cotton industry meet the projected increase in worldwide demand for cotton.

In summary, cotton's future is positive. Cotton use should benefit from consumer demand stemming from favorable economic growth prospects and because of research. On the production side, global output should continue to provide an adequate supply for mill demand. Finally, cotton, one of the most important textile fibers and one of the world's important oilseed crops, should continue to be recognized as a significant commodity in world trade and the consumption of this important fiber, food, and feed crop will continue to grow but at a slower rate than synthetic fibers.

10 Environmental, Workplace, and Consumer Considerations

CONTENTS

10.1 ENVIRONMENT

10.1.1 ENVIRONMENTAL STEWARDSHIP

Cotton, as grown in the United States and most other countries, is an environmentally responsibly produced and managed product [47,627]. Cotton production faces many challenges. Cotton can be affected by insects [628], weeds [629], diseases [630], nematodes [631], and mycotoxins [632]. The newest and latest tested technology is used to produce the crop [9,47]. Modern cotton production minimizes soil erosion by using conservation tillage and other practices, nutrient loss with nutrient management programs, and ground water contamination. At least 90% of the U.S. cotton production uses crop protection products with a wide array of integrated pest management (IPM) programs as well as computer programs and other technologies (e.g., global positioning systems) to apply only the crop protection products that are needed and only where they are needed. A biocontrol (competitive exclusion) method for managing aflatoxin (a mycotoxin that can be a serious food safety hazard on cottonseed) on cotton has been developed for use in Arizona, Texas, and California [633].

10.1.2 EMISSIONS TO THE ENVIRONMENT

The cotton fiber does not contribute anything that causes hazardous air emissions from textile operations processing cotton [9,47]. Cotton production and ginning can be sources of particulate matter (PM) emissions that are regulated by the U.S. EPA. Neither is a major source under EPA regulations. Cotton production can also be a minor source of volatile organic chemicals that are precursors of ozone and of oxides of nitrogen.

For environmental effluent and solid fiber waste concerns and processing of all kinds, it is beneficial for textile mills to know the concentrations of noncellulosic constituents of cotton fiber and what is leachable and removable from the fiber. Then mills know how to handle their water and fiber waste and how to dye and finish the cotton [47].

10.2 WORKPLACE

Workers handle and process cotton through many work operations from harvesting and ginning through yarn and fabric manufacturing.

10.2.1 INHALATION OR RESPIRATORY DISEASE

Inhalation of cotton-related dust, generated during the textile manufacturing operations where cotton fiber is converted into yarn and fabric, has been shown to cause an occupational lung disease, byssinosis, in a small number of textile workers [634,635]. The U.S. Occupational Safety and Health Association (OSHA) regulations for cotton dust apply to textile processing and weaving but do not apply to handling or processing of woven or knitted materials, harvesting, ginning, warehousing, classing or merchandizing, or knitting operations. The OSHA permissible exposure limits (PELs) for cotton dust measured as an 8 h time-weighted-average with the vertical elutriator cotton dust sampler are as follows:

1. Yarn manufacturing, 200 $\mu g/m^3$
2. Textile mill waste house, 500 $\mu g/m^3$
3. Slashing and weaving, 750 $\mu g/m^3$
4. Waste processing (waste recycling and garneting), 1000 $\mu g/m^3$ [636]

It usually takes 15 to 20 years of exposure to higher levels of dust (above 0.5 to 1.0 mg/m^3) for workers to become reactors. Cotton dust, an airborne PM released into the atmosphere as cotton, is handled or processed in textile processing and is a heterogeneous, complex mixture of botanical trash, soil, and microbiological material (i.e., bacteria and fungi), which varies in composition and biological activity [637]. The etiological agent and pathogenesis of byssinosis are not known [638–640]. However, control studies in experimental cardrooms suggest that, in today's world, appropriate engineering controls in cotton textile processing areas, along with work practices, medical surveillance, and personal protective equipment for the most part, can eliminate incidence of workers' reaction to cotton dust [641]. Cotton plant trash associated with the fiber and the endotoxin from Gram-negative bacteria on the fiber, plant trash, and soil are thought to be the causative or to contain the causative associated with workers' reaction to dust [634,635]. The cotton fiber, which is mainly cellulose, is not the causative, because cellulose is an inert dust that does not cause respiratory disease. In fact, cellulose powder has been used as an inert control dust in human exposure studies.

A mild water washing of cotton by batch kier washing systems [642,643] and continuous batt systems [107] reduces the residual level of endotoxin in both lint and airborne dust to below levels associated with a zero percentage change in acute reduction in pulmonary function as measured by forced expiratory volume in 1 sec (FEV_1). Levels of endotoxin from Gram-negative bacteria generated during the processing of cotton are associated with the occupational respiratory disease that affects some textile workers [634,635,643]. Washed cotton is determined by levels of potassium and water-soluble reducing substances (WSRS) in the washed lint [107,642,643]. The OSHA has accepted several mild washing systems as qualifying as washed cotton exemptions under the cotton dust standard [643,644]:

1. Mild washing by the continuous batt system or a rayon rinse system is with water containing a wetting agent, at not less than 60°C, with water-to-fiber ratio not less than 40:1, with bacterial levels in the wash water controlled to limit bacterial contamination of the cotton.
2. The batch kier washing system is with water containing a wetting agent, with a minimum of one wash cycle followed by two rinse cycles for each batch, using fresh water in each cycle, and with bacterial levels in the wash water controlled to limit bacterial contamination of the cotton.
   a. For low temperature, at not less than 60°C, with water-to-fiber ratio not less than 40:1; or
   b. For high temperature, at not less than 93°C, with a water-to-fiber ratio not less than 15:1.

10.2.2 SKIN IRRITATION OR DERMATITIS

Handling or processing conventional U.S. cotton does not cause skin irritation or dermatitis. Cellulose is essentially an inert substance and nothing on the fiber surface is known that could cause dermatitis problems. However, it is remotely possible that some very atypical cottons that have been treated with substances that are not approved for use or that are off-grade and perhaps are highly microbiologically damaged might cause skin irritation. These rare atypical cottons should be evaluated on a case-by-case basis, if they are to be used in a conventional way.

10.2.3 FORMALDEHYDE

Cotton dyeing and finishing operations can expose workers to formaldehyde in concentrations that exceed the U.S. OSHA workplace PEL for formaldehyde of 0.75 ppm of air as an 8 h time weighted average concentration and 2 ppm of air as a 15-min short-term exposure limit (STEL) [645].

Formaldehyde, a component of resins used to impart easy care and durable press and other properties to cotton fabrics, was classified in 1987 by U.S. EPA as a probable human carcinogen (an animal carcinogen and limited evidence that it is a carcinogen in humans) under conditions of unusually high or prolonged exposure [646]. Since then, studies of industrial workers have suggested that formaldehyde is associated with nasopharyngeal cancer and possibly leukemia. In June 2004, the International Agency for Research on Cancer (IARC) reclassified formaldehyde as a known human carcinogen [647]. Sensory irritation of the mucous membranes of the eyes and the respiratory tract, and cellular changes in the nasal cavity are noncancer effects of exposure to low airborne concentrations of formaldehyde.

10.3 CONSUMER

By the time cotton textiles reach the ultimate consumer, there should be nothing known on or extractable from the original cotton fiber that would cause any health concerns to consumers [47]. However, various dyeing and finishing treatments that cotton fabrics are subjected to can leave residues on the fabric or release substances that could cause irritation to consumers, if the treatments are not properly applied.

10.3.1 FORMALDEHYDE

Exposure to formaldehyde from cotton textiles is controlled by the chemical technology on low-emitting formaldehyde resin technology and nonformaldehyde finishes (discussed earlier in Section 6.2 and Section 6.3) and by increased ventilation. It should be noted that in the 1980s, the U.S. Consumer Product Safety Commission (CPSC) studied the effects of various dyeing and finishing treatments, including durable-press finishing of cotton [648,649], and found no acute or chronic health problems of concern to consumers due to exposures to formaldehyde or other finishing chemicals from textiles [650].

In summary, when cotton is grown and processed (including yarn manufacturing and wet processing) in a responsible manner, the production of cotton should not have any adverse effects on the environment (through external emissions, wastewater effluents, and solid wastes), the workers (because of acute or chronic effects), and the consumers (because of acute or chronic effects).

References

1. *Fiber Organon*, 76(7), Fiber Economic Bureau, Inc., Arlington, VA, 2005.
2. Basra, A.S., Ed., *Cotton Fibers, Developmental Biology, Quality Improvement, and Textile Processing*, Food Products Press, The Haworth Press, Binghamton, New York, 1999.
3. Bell, T.M. and Gillham, F.E.M., *The World of Cotton*, ContiCotton, EMR, Washington, D.C., 1989.
4. Chaudhry, M.R. and Guitchounts, A., *Cotton Facts*, International Cotton Advisory Committee (ICAC), Washington, D.C., 2003.
5. Hake, S.J, Kerby, T.A., and Hake, K.D., Eds., *Cotton Production Manual*, Pub. 3352, University of California, Division of Agriculture and Natural Resources, Oakland, CA, 1996.
6. Hamby, D.S., Ed., *The American Cotton Handbook*, 3rd ed., Vols. 1 and 2, Interscience, New York, 1965.
7. Cook, J.G., Natural fibres, in *Handbook of Textile Fibres*, 4th ed., Vol. 1, Merrow, Watford, England, 1968.
8. Wakelyn, P.J., Cotton, in *Encyclopedia of Polymer Science and Technology*, 3rd ed., Vol. 5, John Wiley & Sons, New York, 2004, pp. 721–759.
9. Wakelyn, P.J., Bertoniere, N.R., French, A.D., Thibodeaux, D.P., Tripplett, B.A., Goynes, W.R., Jr., Hughs, S.E., Knowlton, J.L., Norman, B.M., and Lanclos, D.K., Cotton, in *Kirk–Othmer Encyclopedia of Chemical Technology*, 5th ed., Vol. 8, John Wiley & Sons, New York, 2004, pp. 1–40.
10. Gillham, F.E.M., Bell, T.E., Ryan, P.D., and Gilson, S.R., *Cotton from Field to Fabric*, Vol. 1, Tom Bell Associates, Falls Church, VA, 1993.
11. Kohel, R.J. and Lewis, C.F., Eds., *Cotton*, Agronomy Monograph No. 24, American Society of Agronomy, Crop Sciences Society of America and Soil Science Society of America, Madison, WI, 1984.
12. French, A.D. and Bertoniere, N.R., Cellulose, in *Kirk–Othmer Encyclopedia of Chemical Technology*, 4th ed., Vol. 5, Interscience, New York, 1993, pp. 476–496.
13. Ward, K., Jr., Ed., *Chemistry and Chemical Technology of Cotton*, Interscience, New York, 1955.
14. Bikales, N.M. and Segal, L., Eds., *Cellulose and Cellulose Derivatives*, Parts IV and V, Wiley Interscience, New York, 1971.
15. Ward, K., Jr., and Morak, A.J., *Chemical Reactions of Polymers*, Fettes, E.M., Ed., Interscience, New York, 1964, pp. 321–365.
16. Nevell, T.P. and Zeronian, S.H., Eds., *Cellulose Chemistry and Its Applications*, Ellis Horwood Ltd., Chichester, 1985, pp. 1–551.
17. Bertoniere, N.R., The chemical nature of cellulose, Proceedings of the Beltwide Cotton Conferences, National Cotton Council, Memphis, TN, 1993, pp. 1413–1416.
18. French, A.D., Molecular arrangements in cellulose, Proceedings of the Beltwide Cotton Conferences, National Cotton Council, Memphis, TN, 1993, pp. 1417–1420.
19. Munro, J.M., *Cotton*, 2nd ed., Longman Scientific and Technical, Essex, England and John Wiley & Sons, New York, 1987.
20. Rayburn, S.T., Bitton, R., and Keen, E., *National Variety Test: Yield, Boll, Seed, Spinning and Data*, USDA, ARS, Stoneville, MS, 1992.
21. Fitt, G.P., Wakelyn, P.J., Stewart, J.M., Roupakias, D., Pages, J., Giband, M., Zafar, Y., Hake, K., and James, C., Report of the Second Expert Panel on Biotechnology in Cotton, International Cotton Advisory Committee (ICAC), Washington, D.C., November 2004.

22. Wakelyn, P.J., May, O.L., and Menchey, E.K., Cotton and biotechnology, in *Handbook of Plant Biotechnology*, Christou, P. and Klee, H., Eds., John Wiley & Sons, Chichester, West Sussex, UK, 2004, chap. 57, pp. 1117–1131.

23. Vreeland, J.M., Jr., Naturally colored and organically grown cottons: anthropological and historical perspectives, Proceedings of the Beltwide Cotton Conferences, National Cotton Council, Memphis, TN, 1993, pp. 1533–1536.

24. Vreeland, J.M., Jr., *Sci. Am.*, 280, 112, 1999.

25. Wakelyn, P.J. and Gordan, M.B., *Text. Horizons*, 15(1), 36–38, 1995.

26. Kimmel, L.B. and Day, M.P., *AATCC Rev.*, 1(10), 32, 2001.

27. Ryser, U., Cotton fiber initiation and histodifferentiation, in *Cotton Fibers, Developmental Biology, Quality Improvement, and Textile Processing*, Basra, A.S., Ed., Haworth Press, Binghamton, New York, 1999, chap. 1, pp. 21–29.

28. Ryser, U., Meier, H., and Holloway, P.J., *Protoplasma*, 117, 196, 1983.

29. Schmutz, A., Jenny, T., Amrhein, N., and Ryser, U., *Planta*, 189, 453, 1993.

30. The National Organic Program, available at http://www.ams.usda.gov/nop/indexIE.htm, Agricultural Marketing Service, USDA.

31. Chaudhry, M.R., *The ICAC Recorder*, Vol. XVI, No. 4, December 1998; Chaudhry, M.R., *The ICAC Recorder*, Vol. XXI, No. 1, March 2003.

32. Guillou, G. and Scharpé, A., *Organic Farming—Guide to Community Rules*, Directorate General for Agriculture, European Commission, ISBN 92-894-0363-2, 2000.

33. Myers, D. and Stolton, S., *Organic Cotton—from Field to Final Product*, Intermediate Technology Publications, Intermediate Technology Development Group, 103–105 Southhampton Row, London, WC1B 4HH, UK., 1999.

34. Wakelyn, P.J. and Chaudtry, M.R., Organic Cotton, in *Cotton: Science and Technology*, Gordon, S. and Heich, L., Eds., Woodhead Publishing Limited, UK, 2006.

35. Kyrgyzstan: organic cotton tested in the south, Reuters Alert Net, Dec 28, 2004. (http://www.alertnet.org/).

36. Marquardt, S., Organic cotton: production and market trends in the United States and Canada—2001 and 2002, Proceedings of the 2003 Beltwide Cotton Conference, National Cotton Council, Memphis, TN, 2003, pp. 362–366.

37. U.S. Organic Cotton Production Drops Despite Increasing Sales of Organic Cotton Products, Organic Trade Association, News Release, December 28, 2004. (http://www.ota.com).

38. Organic Exchange, http://organicexchange.org, 2005.

39. Variation in yields among countries, *ICAC Recorder*, Vol. XXII, No. 3, September 2004, pp. 3–7.

40. U.S. Department of Agriculture, Foreign Agriculture Service, June 2005. (http://www.fas.usda.gov/psd/).

41. Anthony, W.S. and Mayfield, W.D., Eds., *Cotton Ginners Handbook*, U.S. Dept. of Agriculture, *Agricultural Handbook 503*, Washington, D.C., 1994.

42. Wakelyn, P.J., Thompson, D.W., Norman, B.M., Nevius, C.B., and Findley, D.S., *Cotton Gin and Oil Mill Press*, 106(8), 5, 2005.

43. O'Connor, R.T., *Instrumental Analysis of Cotton Cellulose and Modified Cotton Cellulose*, Marcel Dekker, New York, 1972.

44. Berlin, J.D. and Watson, M., Fine structural differentiation of lint and fuzz fibers, Proceedings of the Beltwide Cotton Production Research Conferences, National Cotton Council of America, Memphis, TN, 1974, p. 52.

45. Supak, J.R. and Snipes, Eds., *Cotton Harvest Management: Use and Influence of Harvest Aids*, The Cotton Foundation Reference Book Series, No. 5, The Cotton Foundation, Memphis, TN, 2001; Chaudhry, M.R., Harvesting and ginning of cotton in the world, Proceedings of the Beltwide Cotton Production Research Conferences, Vol. 2, National Cotton Council of America, Memphis, TN, 1997, pp. 1617–1619.

46. ISO 8115, Cotton bales—dimensions and density, International Organization for Standardization, Switzerland, 1986.

47. Wakelyn, P.J., Cotton: environmental concerns and product safety, Proceedings of the 22nd International Cotton Conferences, Bremen, Harig, H. and Heap, S.A., Eds., Faserinstitut Bremen eV, Bremen, Germany, 1994, pp. 287–305.
48. Beasley, C.A. and Ting, I.P., *Am. J. Bot.*, 61, 188, 1974.
49. Stewart, J.McD., *Am. J. Bot.*, 62, 723, 1975.
50. Temming, H., *Linters: Technical Information on Cotton Cellulose*, Glückstadt, West Germany, 192, 1973.
51. Gamble, G., *J. Agric. Food Chem.*, 51, 7995, 2003.
52. Seagull, R.W., *Tip Growth in Plant and Fungal Cells*, Heath, I.B., Ed., Academic Press, San Diego, CA, 1990, pp. 261–284.
53. Tiwari, S.C. and Wilkins, T.A., *Can. J. Bot.*, 73, 746, 1995.
54. Tokumoto, H., Wakabayashi, K., Kamisaka, S., and Hoson, T., *Plant Cell Physiol.*, 43, 411, 2002.
55. Tokumoto, H., Wakabayashi, K., Kamisaka, S., and Hoson, T., *J. Plant Physiol.*, 160, 1411, 2003.
56. Reiter, W.D., *Curr. Opin. Plant Biol.*, 5, 536, 2002.
57. Scheible, W.R. and Pauly, M., *Curr. Opin. Plant Biol.*, 7, 285, 2004.
58. Arpat, A.B., Waugh, M., Sullivan, J.P., Gonzales, M., Frisch, D., Main, D., Wood, T., Leslie, A., Wing, R.A., and Wilkins, T.A., *Plant Mol. Biol.*, 54, 911, 2004.
59. H.R. Huwyler, G. Franz, and H. Meier, *Plant Sci. Lett.*, 12, 55, 1978.
60. Huwyler, H.R., Franz, G., and Meier, H., *Planta*, 46, 635, 1979.
61. Maltby, D., Carpita, N.C., Montezinos, D., Kulow, C., and Delmer, D.P., *Plant Physiol.*, 63, 1158, 1979.
62. Waterkeyn, L., *Protoplasma*, 106, 49, 1981.
63. Cui, X., Shin, H., Song, C., Laosinchai, W., Amano, Y., and Brown, R.M., Jr., *Planta.*, 213, 223, 2001.
64. Wakeham, H. and Spicer, N., *Text. Res. J.*, 25, 585, 1955.
65. Hearle, J.W.W. and Sparrow, J.T., *Text. Res. J.*, 41, 736, 1971.
66. Moharir, A.V., van Langenhove, L., van Nimmen, E., Louwagie, J., and Kiekens, P., *J. Appl. Polym. Sci.*, 72, 269, 1999.
67. Seagull, R.W., *J. Cell Sci.*, 101, 561, 1992.
68. Dixon, D.C., Seagull, R.W., and Triplett, B.A., *Plant Physiol.*, 105, 1347, 1994.
69. Andersland, J.M., Dixon, D.C., Seagull, R.W., and Triplett, B.A., *In Vitro Cell Develop. Biol. Plant*, 34, 173, 1998.
70. Whittaker, D.J. and Triplett, B.A., *Plant Physiol.*, 121, 181, 1999.
71. Dixon, D.C., Meredith, W.R., Jr., and Triplett, B.A., *Int. J. Plant Sci.*, 161, 63, 2000.
72. Andersland, J.M. and Triplett, B.A., *Plant Physiol. Biochem.*, 38, 193, 2000.
73. Delmer, D.P., Heiniger, U., and Kulow, C., *Plant Physiol.*, 59, 713, 1977.
74. Amor, Y., Haigler, C.H., Wainscott, M., Johnson, S., and Delmer, D.P., *Proc. Natl. Acad. Sci. U.S.A.*, 92, 9353, 1995.
75. Haigler, C.H., Ivanova-Datcheva, M., Hogan, P.S., Salnikov, V.V., Hwang, S., Martin, K., and Delmer, D.P., *Plant Mol. Biol.*, 47, 29, 2001.
76. Peng, L., Kawagoe, Y., Hogan, P., and Delmer, D.P., *Science*, 295, 147, 2002.
77. Schrick, K., Fujioka, S., Takatsuto, S., Stierhof, Y.D., Stransky, H., Yoshida, S., and Jurgens, G., *Plant J.*, 38, 227, 2004.
78. Meuller, S.C. and Brown, R.M., Jr., *J. Cell Biol.*, 84, 315, 1980.
79. Haigler, C.H. and Brown, R.M., Jr., *Protoplasma*, 134, 111, 1986.
80. Hayashi, T., Read, S.M., Bussell, J., Thelen, M.T., Lin, F.-C., Brown, R.M., Jr., and Delmer, D.P., *Plant Physiol.*, 83, 1054, 1987.
81. Okuda, R., Li, L., Kudlicka, K., Kuga, S., and Brown, R.M, Jr., *Plant Physiol.*, 101, 1131, 1993.
82. Lai-Kee-Him, J., Chanzy, H., Müller, M., Puteaux, J.-L., Imai, T., and Bulone, V., *J. Biol. Chem.*, 277, 36931, 2002.
83. Pear, J.R., Kawagoe, Y., Schreckengost, W.E., Delmer, D.P., and Stalker, D.M., *Proc. Natl. Acad. Sci. U.S.A.*, 93, 12637, 1996.
84. Kimura, S., Laosinchai, W., Itoh, T., Cui, X., Linder, C.R., and Brown, R.M., Jr., *Plant Cell*, 11, 2075, 1999.

85. Richmond, T. and Somerville, C.R., *Plant Mol. Biol.*, 47, 131, 2001.

86. Taylor, N.G., Howells, R.M., Huttly, A.K., Vickers, K., and Turner, S.R., *Proc. Natl. Acad. Sci. U.S.A.*, 100, 1450, 2003.

87. Arioli, T., Peng, L., Betzner, A.S., Burn, J., Wittke, W., Herth, W., Camilleri, C., Höfte, H., Plazinski, J., Birch, R., Cork, A., Glover, J., Redmond, J., and Williamson, R.E., *Science*, 279, 717, 1998.

88. Schieble, W.-R., Eshed, R., Richmond, T., Delmer, D., and Somerville, C., *Proc. Natl. Acad. Sci. U.S.A.*, 98, 10079, 2001.

89. Holland, N., Holland, D., Helentjaris, T., Dhugga, K.S., Xoconostle-Cazares, B., and Delmer, D.P., *Plant Physiol.*, 123, 1313, 2000.

90. Taylor, N.G., Laurie, S., and Turner, S.R., *Plant Cell*, 12, 2529, 2000.

91. Kim, H.J. and Triplett, B.A., Regulation of gene expression in the transition from cell elongation to secondary wall formation in cotton fiber, Proceedings of the Beltwide Cotton Conferences, National Cotton Council, Memphis, TN, 2005, 1043.

92. Adams, K.L., Cronn, R., Percifield, R., and Wendel, J.F., *Proc. Natl. Acad. Sci. U.S.A.*, 100, 4649, 2003.

93. Marx-Figini, M., *Nature*, 210, 755, 1966.

94. Timpa, J.D. and Triplett, B.A., *Planta*, 189, 101, 1993.

95. Haigler, C.H., Zhang, D., and Wilkerson, C.G., *Physiol. Plant.*, 124, 285, 2005.

96. Tripp, V.W., Moore, A.T, and Rollins, M.L., *Text. Res. J.*, 21, 886, 1951.

97. Guthrie, J.D., The chemistry of lint cotton, in *Chemistry and Chemical Technology of Cotton*, Ward, K., Jr., Ed., Interscience, New York, 1955, p. 2.

98. McCall, E.R. and Jurgens, J.F., *Text. Res. J.*, 21, 19, 1951.

99. Warwicker, J.O., Jeffries, R., Colbran, R.L., and Robinson, R.N., A Review of the Literature on the Effect of Caustic Soda and Other Swelling Agents on the Fine Structure of Cotton, Pamphlet No. 93, Shirley Institute, Manchester, UK, 1966.

100. Rollins, M.L., *For. Prod. J.*, 18(2), 91, 1968.

101. Catlett, M.S., Giuffria, R., Moore, A.T., and Rollins, M.L., *Text. Res. J.*, 21, 880, 1951.

102. Tripp, V.W. and Rollins, M.L., *Anal. Chem.*, 24, 1721, 1952.

103. Wakelyn, P.J., *Text. Res. J.*, 45, 418, 1975.

104. Goldthwaite, C.F., Kettering, J.H., and Guthrie, J.D., Chemical properties of cotton fiber, in *Matthews' Textile Fibers*, 5th ed., Mauersberger, H.P., Ed., Wiley Interscience, New York, 1947, pp. 266–304 [see also 6th ed., 1954].

105. Ferreti, R.J., Merola, G.V., Marsh, P.B., and Simpson, M.E., *Cotton Grow. Rev.*, 52, 136, 1975.

106. Scmutz, A., Buchala, A.J., and Ryser, U., *Plant Physiol.*, 110, 403, 1996.

107. Wakelyn, P.J., Jacobs, R.R., and Kirk, I.W., Eds., *Washed Cotton: Washing Techniques, Processing Characteristics, and Health Effects*, USDA, Washington, D.C., 1986.

108. Vaughn, K.C. and Turley, R.B., *Protoplasma*, 209, 226, 1999.

109. Brushwood, D.E. and Perkins, H.H., Jr., Variations in cotton insect honeydew composition and the related effects on test methods and processing quality, Proceedings of the Beltwide Cotton Conferences, National Cotton Council, Memphis, TN, 1995, p. 1178.

110. Hector, D.J. and Hodkinson, I.D., Stickiness in cotton, *ICAC Review Articles on Cotton Production Research*, No. 2, International Cotton Advisory Committee, 1901 Pennsylvania Avenue, Washington, D.C., 1989.

111. Perkins, H.H., Jr., A survey of sugar and sticky cotton test methods, Proceedings of the Beltwide Cotton Conferences, National Cotton Council, Memphis, TN, 1993, 1136.

112. Brushwood, D.E. and Perkins, H.H., Jr., Characterization of sugar from honeydew contaminated and normal cottons, Proceedings of the Beltwide Cotton Conferences, National Cotton Council, Memphis, TN, 1994, p. 1408.

113. Bourely, J., *Cotton Fibers Tropicale*, 35(2), 189, 1980.

114. Cheung, P.S.R., Roberts, C.W., and Perkins, H.H., Jr., *Text. Res. J.*, 50, 55, 1980.

115. Hendrix, D.L., Wei, Y.A., and Leggett, J.E., *Comp. Biochem. Physiol.*, 101B, 23, 1992.

116. Tarczyski, M.C., Byrne, D.N., and Miller, W.B., *Plant Physiol.*, 98, 753, 1992.

117. Bates, R.B., Byrne, D.N., Kane, V.V., Miller, W.B., and Taylor, S.R., *Carbohydr. Res.*, 201, 342, 1990.

118. Byrne, D.N. and Miller, W.B., *J. Insect Physiol.*, 36, 433, 1990.

119. Roberts, C.W., Cheung, P.S.R., and Perkins, H.H., Jr., *Text. Res. J.*, 48, 91, 1978.

120. Brushwood, D.E. and Perkins, H.H., Jr., *Text. Chem. Color.*, 26(3), 32, 1994.
121. Matsumura, R.T., Ashbaugh, L., James, T., Carvacho, O., and Flocchini, R., Size distribution of PM-10 soil dust emissions from harvesting crops, Proceedings of the International Conferences on Air Pollution from Agricultural Operations, Kansas City, MO, February 7–9, 1996.
122. Columbus E.P. and Morris, N.M., *Trans. Am. Soc. Ag. Eng.*, 27, 546, 1984.
123. Perkins, H.H., Jr. and Brushwood, D.E., *Text. Chem. Color.*, 23(2), 26, 1991.
124. Greenwood, P.F., *Text.*, 2, 23, 1993.
125. Cook, F.C., *Textile World*, May, 84, 1991.
126. Rucker, J.W., Freeman, H.S., and Hsu, W.-N, *Text. Chem. Color.*, 24(9), 66, 1992.
127. Wakelyn, P.J., Supak, J., Carter, F.C., and Roberts, B., Public and environmental issues, in *Cotton Harvest Management: Use and Infuence of Harvest Aids*, The Cotton Foundation Reference Book Series, No. 5, Supak, J.R. and Snipes, C.E., Eds., The Cotton Foundation, Memphis, TN, 2001, chap. 10, pp. 275–302.
128. Turbak, A.F., Hammer, R.B., Davies, R.E., and Hergert, H.L., *Chem. Tech.*, 10, 51, 1980.
129. Hudson, S.M. and Cuculo, J.A., *J. Macromol. Sci.*, C18, 1, 1980.
130. Turbak, A.F., Newer cellulose solvent systems, Proceedings of the 1983 International Dissolving and Specialty Pulps Conferences, *TAPPI*, Atlanta, U.S., 1983.
131. Johnson, D.C., Solvents for cellulose, in *Cellulose Chemistry and Its Applications*, Nevell, T.P. and Zeronian, S.H., Eds., Ellis Horwood Ltd., Chichester, 1985, chap. 7, pp. 181–201.
132. Treiber, E.E., Formation of fibers from cellulose solution, in *Cellulose Chemistry and Its Applications*, Nevell, T.P. and Zeronian, S.H., Eds., Ellis Horwood Ltd., 1985, chap. 18, pp. 455–479.
133. Johnson, D.C., Nicholson, M.D., and Haigh, F.L., *Appl. Polym. Symp.*, 28, 931, 1976.
134. Jayme, G., Investigations of solutions: new solvents, in *High Polymers*, Vol. V, *Cellulose and Cellulose Derivatives*, Part IV, Bikales, N.M. and Segal, L., Eds., Wiley Interscience, New York, 1971, p. 381.
135. Timpa, J.D., *J. Agric. Food Chem.*, 39, 270, 1991.
136. Graenacher, C. and Sallmann, R., U.S. Patent 2179181, November 7, 1939.
137. McCorsley, C.C, III, and Varga, J.K., U.S. Patent 4142913, March 6, 1979.
138. McCorsely, C.C., III, U.S. Patent 4246221, January 20, 1981.
139. Augustine, A.V., Hudson, S.M., and Cuculo, J.A., Direct solvents for cellulose, in *Steam Explosion Techniques, Fundamentals and Industrial Applications*, Focher, B., Marzetti, A., and Crescenzi, V., Eds., Gordon & Breach, Philadelphia, PA, 1991, p. 251.
140. Rules and regulations under the textile fiber products identification act, Federal Register, 60, pp. 62352–62354, December 6, 1995.
141. Mitzutani, C., Wartell, L.H., Bertoniere, N.R., and French, A.D., unpublished data, 1995.
142. Westermark, U. and Gustafsson, K., *Holzforschung*, 48, 146, 1994.
143. Sjoholm, E., Gustafsson, K., Pettersson, B., and Colmsjo, A., *Carbohydr. Polym.*, 32, 57, 1997.
144. Dupont, A.L., *Polymer*, 44, 4117, 2003.
145. Potthast, A., Rosenau, T., Sixta, H., and Kosma, P., *Tetrahedron Lett.*, 43, 7757, 2002.
146. Strlic, M. and Kolar, J., *J. Biochem. Biophys. Methods*, 56, 265, 2003.
147. Takaragi, A., Minoda, M., Miyamote, T., Liu, H.Q., and Zhang, L.N., *Cellulose*, 6, 93, 1999.
148. Rousselle, M.A., *Text. Res. J.*, 72(2), 131, 2002.
149. Yanagisawa, M., Shibata, I., and Isogai, A., *Cellulose*, 11, 169, 2004.
150. Kamide, K. and Okajima, K., U.S. Patent 4634470, 1987.
151. Isogai, A. and Atalla, R.H., *Cellulose*, 5, 309, 1998.
152. Swatloski, R.P., Spear, S.K., Holbrey, J.D., and Rogers, R.D., *J. Am. Chem. Soc.*, 124, 4974, 2002.
153. Broughton, R., Wang, W., Farag, R., Rogers, R., and Swatloski, R., Cellulose Fibers Extruded from Ionic Liquids, paper presented at INTC 2004, Toronto, Canada, September 20–23, 2004.
154. Krassig, H. and Kappner, W., *Makromol. Chem.*, 44–46, 1, 1961.
155. Nishiyama, Y., Kim, U.-J., Kim, D.-Y., Katsumata, K.S., May, R.P., and Langan, P., *Biomacromolecules*, 4, 1013, 2003.
156. Ziderman, I.I. and Perel, J., *J. Macromol. Sci. Phys.*, B24, 181, 1985–1986.
157. Hessler, L.E., Merola, G.V., and Berkley, E.E., *Text. Res. J.*, 18, 628, 1948.

158. Marx-Figini, M., *Cellulose and Other Natural Polymer Systems: Biogenesis, Structure, and Degradation*, Brown, R.M., Ed., Plenum Press, New York, 1982, p. 243.

159. McCormick, C.L., Callais, P.A., and Hutchinson, B.H., *Macromolecules*, 18, 2394, 1985.

160. Turbak, A.F., *Wood and Agricultural Residues*, Soltes, E.J., Ed., Academic Press, New York, 1983, 87.

161. Kennedy, J.F., Rivera, Z.A., White, C.A., Lloyd, L.L., and Warner, F.P., *Cellul. Chem. Technol.*, 24, 319, 1990.

162. Dawsey, T.R. and McCormick, C.L., *Rev. Macromol. Chem. Phys.*, C30, 403, 1990.

163. Meinert, M.C. and Delmer, D.P., *Plant Physiol.*, 59, 1088, 1977.

164. Timpa, J.D. and Ramey, H.H., *Text. Res. J.*, 59, 661, 1989.

165. Timpa, J.D. and Ramey, H.H., *Text. Res. J.*, 64, 557, 1994.

166. Timpa, J.D. and Wanjura, D.F., *Cellulose and Wood: Chemistry and Technology*, Schuerch, C., Ed., John Wiley & Sons, New York, 1989, p. 1145.

167. Segal, L. and Timpa, J.D., *Text. Res. J.*, 43, 185, 1973.

168. Rousselle, M.A. and Howley, P.S., *Text. Res. J.*, 68, 606, 1998.

169. Morgado, J., Cavaco-Paulo, A., and Rousselle, M.A., *Text. Res. J.*, 70, 696, 2000.

170. French, A.D. and Johnson, G.P., *Cellulose*, 11, 5, 2004.

171. Brant, D.A. and Christ, M.D., Realistic conformational modeling of carbohydrates: applications and limitations in the context of carbohydrate-high polymers, in *Computer Modeling of Carbohydrate Molecules*, French, A.D. and Brady, J.W., Eds., ACS Symposium Series, No. 430, American Chemical Society, Washington, D.C., 1990, p. 42.

172. Meader, D., Atkins, E.D.T., and Happey, F., *Polymer*, 19, 1371, 1978.

173. Zugenmaier, P., Konformations- und packungsanalyse von polysacchariden, in *Polysaccharide*, Burchard, W., Ed., Springer–Verlag, Berlin, 1985, p. 271.

174. Whitaker, P.M., Nieduzynski, I.A., and Atkins, E.D.T., *Polymer*, 15, 125, 1974.

175. Sarko, A., Nishimura, H., and Okano, T., Crystalline alkali–cellulose complexes as intermediates during mercerization, in *The Structures of Cellulose*, Atalla, R.H., Ed., ACS Symposium Series, No. 340, American Chemical Society, Washington, D.C., 1987, p. 169.

176. Jones, D.W., *J. Polym. Sci.*, 42, 173, 1960.

177. Allen, F.H., *Acta Crystallogr.*, Sect. B, 58, 380, 2002.

178. Berman, H.M., Battistuz, T., Bhat, T.N., Bluhm, W.F., Bourne, P.E., Burkhardt, K., Feng, Z., Gilliland, G.L., Iype, L., Jain, S., Fagan, P., Marvin, J., Padilla, D., Ravichandran, V., Schneider, B., Thanki, N., Weissig, H., Westbrook, J.D., and Zardecki, C., *Acta Crystallogr.*, Sect. D, 58, 899, 2002.

179. Chu, S.S.C. and Jeffrey, G.A., *Acta Crystallogr.*, Sect. B, 24, 830, 1968.

180. Peralta-Inga, Z., Johnson, G.P., Dowd, M.K., Rendleman, J.A., Stevens, E.D., and French, A.D., *Carbohydr. Res.*, 337, 851, 2002.

181. Rencurosi, A., Röhrling, J., Pauli, J., Potthast, A., Jäger, C., Pérez, S., Kosma, P., and Imberty, A., *Angew. Chem., Int. Ed. Engl.*, 41, 4277, 2002.

182. Ernst, A. and Vasella, A., *Helv. Chim. Acta.*, 79, 1279, 1996.

183. Leung, F. and Marchessault, R.H., *Can. J. Chem.*, 51, 1215, 1973.

184. Perez, S. and Brisse, F., *Acta Crystallogr.*, Sect. B, 33, 2578, 1977.

185. Burkert, U. and Allinger, N.L., Molecular mechanics, in *ACS Monograph 177*, American Chemical Society, Washington, D.C., 1982.

186. Hardy, B.J. and Sarko, A., *J. Comput. Chem.*, 14, 848, 1993.

187. Schmidt, R., Trojan, C., Tasaki, K., and Brady, J.W., Proceedings of the Second Tricel Symposium on Trichoderma Reesei Celulases and other Hyrdrolayses, Suominen, P. and Reinikainen, T., Eds., Foundation for Biotechnical and Industrial Fermentation Research 8, 1993, p. 41.

188. Strati, G.L., Willett, J.L., and Momany, F.A., *Carbohydr. Res.*, 337, 1833, 2002.

189. French, A.D., Johnson, G.P., Kelterer, A.-M., and Csonka, G.I., *Tetrahedron: Asymmetry*, 16, 577, 2005.

190. Kroon-Batenburg, L.M.J., Kroon, J., Leeflang, B.R., and Vliegenthart, J.F.G., *Carbohydr. Res.*, 245, 21, 1993.

191. Tanaka, F. and Fukui, N., *Cellulose*, 11, 33, 2004.

192. Umemura, M., Yuguchi, Y., and Hirotsu, T., *J. Phys. Chem. A*, 108, 7063, 2004.

193. Hardy, B.J. and Sarko, A., *J. Comput. Chem.*, 14, 831, 1993.
194. Kroon-Batenburg, L.M.J., Kroon, J., and Northolt, J.G., *Polym. Commun.*, 27, 290, 1986.
195. Sturcova, A., His, I., Apperley, D.C., Sugiyama, J., and Jarvis, M.C., *Biomacromolecules*, 5, 1333, 2004.
196. Atalla, R.H., *Comprehensive Natural Products Chemistry*, Barton, D., Nakanishi, K., and Meth-Cohn, O., Eds., Vol. 3, *Carbohydrates and Their Derivatives Including Tannins, Cellulose, and Related Lignins*, Pinto, B.M., Ed., Elsevier, Amsterdam, 1999, p. 169.
197. Klug, H.P. and Alexander, L.E., *X-Ray Diffraction Procedures*, 2nd ed., John Wiley & Sons, New York, 1974, p. 13.
198. Nishiyama, Y., Kuga, S., Wada, M., and Okano, T., *Macromolecules*, 30, 6395, 1997.
199. Paralikar, K.M., Betrabet, S.M., and Bhat, N.V., *J. Appl. Crystallogr.*, 12, 589, 1979.
200. Jeffrey, G.A. and Park, Y.J., *Acta Crystallogr., Sect. B*, 28, 257, 1972.
201. Jeffrey, G.A. and Huang, D.-B., *Carbohydr. Res.*, 222, 47, 1991.
202. Atalla, R.H. and van der Hart, D.L., *Science*, 223, 283, 1984.
203. Sugiyama, J., Vuong, R., and Chanzy, H., *Macromolecules*, 24, 4168, 1991.
204. Imai, T. and Sugiyama, J., *Macromolecules*, 31, 6275, 1998.
205. Atalla, R.H. and van der Hart, D.L., *Solid State Nucl. Magn. Reson.*, 15, 1, 1999.
206. Wickholm, K., Larsson, P.T., and Iversen, T., *Carbohydr. Res.*, 312, 123, 1998.
207. Wada, M., Okano, T., and Sugiyama, J., *J. Wood. Sci.*, 47, 124, 2001.
208. Nishiyama, Y., Sugiyama, J., Chanzy, H., and Langan, P., *J. Am. Chem. Soc.*, 125, 14300, 2003.
209. Nishiyama, Y., Langan, P., and Chanzy, H., *J. Am. Chem. Soc.*, 124, 9074, 2002.
210. Wada, M., Kondo, T., and Itoh, T., *Polym. J.*, 35, 155, 2003.
211. Wada, M., *J. Polym. Sci., Part B: Polym. Phys.*, 40, 1095, 2002.
212. Kuutti, L., Peltonen, J., Pere, J., and Teleman, O., *J. Microsc.*, 178, 1, 1995.
213. Baker, A.A., Helbert, W., Sugiyama, J., and Miles, M., *J. Biophys.*, 79, 1139, 2000.
214. Vietor, R.J., Newman, R.H., Ha, M.-A., Apperley, D.C., and Jarvis, M.C., *Plant J.*, 30, 721, 2002.
215. Newman, R.H. and Davidson, T.C., *Cellulose*, 11, 23, 2004.
216. Sasaki, M., Adschiri, T., and Arai, K., *J. Agric. Food Chem.*, 51, 5376, 2003.
217. Roberts, E.M., Saxena, I.M., and Brown, R.M., *Cellulose and Wood—Chemistry and Technology*, Schuerch, C., Ed., Wiley Interscience, New York, 1989, p. 689.
218. Kuga, S., Takagi, S., and Brown, R.M., *Polymer*, 34, 93, 1993.
219. Hirai, A., Tsuji, M., and Horii, F., *Cellulose*, 9, 105, 2002.
220. Langan, P., Nishiyama, Y., and Chanzy, H., *J. Am. Chem. Soc.*, 121, 9940, 1999.
221. Langan, P., Nishiyama, Y., and Chanzy, H., *Biomacromolecules*, 2, 410, 2001.
222. Gessler, K., Krauss, N., Steiner, T., Betzel, C., Sarko, A., and Saenger, W., *J. Am. Chem. Soc.*, 117, 11397, 1995.
223. Raymond, S., Heyraud, A., Tran Qui, D., Kvick, A., and Chanzy, H., *Macromolecules*, 28, 2096, 1995.
224. Raymond, S., Henrissat, B., Tran Qui, D., Kvick, A., and Chanzy, H., *Carbohydr. Res.*, 277, 209, 1995.
225. Noordik, J.H., Beurskens, P.T., Bennema, P., Visser, R.A., and Gould, R.O., *Z. Kristallogr.*, 168, 59, 1984.
226. Wada, M., Heux, L., and Sugiyama, J., *Biomacromolecules*, 5, 1385, 2004.
227. Wada, M., Heux, L., Isogai, A., Nishiyama, Y., Chanzy, H., and Sugiyama, J., *Macromolecules*, 22, 3168, 2001.
228. Wada, M., Chanzy, H., Nishiyama, Y., and Langan, P., *Macromolecules*, 37, 8548, 2004.
229. Chanzy, H., Imada, K., Mollard, A., Vuong, R., and Barnoud, F., *Protoplasma*, 100, 303, 1979.
230. Kondo, T., Togawa, E., and Brown, R.M., Jr., *Biomacromolecules*, 2, 1324, 2001.
231. Atalla, R.H. and Nagel, S.C., *Science*, 185, 522, 1974.
232. Whimore, R.E. and Atalla, R.H., *Int. J. Biol. Macromol.*, 7, 182, 1985.
233. Atalla, R.H., Ellis, J.D., and Schroeder, L.R., *J. Wood Chem. Technol.*, 4, 465, 1984.
234. Shibazaki, H., Kuga, S., and Okano, T., *Cellulose*, 4, 75, 1997.
235. Dinand, E., Vignon, M., Chanzy, H., and Heux, L., *Cellulose*, 9, 7, 2002.
236. Blackwell, J., Kurz, D., Su, M.-Y., and Lee, D.M., X-ray studies of the structure of cellulose complexes, in *The Structures of Cellulose*, Atalla, R.H., Ed., ACS Symposium Series, No. 340, American Chemical Society, Washington, D.C., 1987, p. 199.

237. Lee, D.M. and Blackwell, J., *Biopolymers*, 20, 2165, 1981.

238. Stipanovic, A.J. and Sarko, A., *Polymer*, 19, 3, 1978.

239. Sikorski, P., Wada, M., Heux, L., Shintani, H., and Stokke, B.T., *Macromolecules*, 37, 4547, 2004.

240. Kuga, S. and Brown, R.M., *Carbohydr. Res.*, 180, 345, 1988.

241. Koyama, M., Helbert, W., Imai, T., Sugiyama, J., and Henrissat, B., *Proc. Natl. Acad. Sci. U.S.A.*, 94, 9091, 1997.

242. Kuga, S. and Brown, R.M., Jr., *J. Electron Microsc. Tech.*, 6, 349, 1987.

243. Kuga, S. and Brown, R.M., Jr., *Polymer Commun.*, 28, 311, 1987.

244. Lin, J.S., Tang, M.-Y., and Fellers, J.F., Fractal analysis of cotton cellulose as characterized by small-angle x-ray scattering, in *The Structures of Cellulose*, Atalla, R.H., Ed., ACS Symposium Series, No. 340, American Chemical Society, Washington, D.C., 1987, p. 234.

245. Statton, W.O. and Godard, G.M., *J. Appl. Phys.*, 28, 1111, 1957.

246. Klug, H.P. and Alexander, L.E., *X-Ray Diffraction Procedures*, John Wiley & Sons, New York, 1954, chap. 9.

247. Segal, L., Creely, J.J., Martin, A.E., Jr., and Conrad, C.M., *Text. Res. J.*, 29, 786, 1959.

248. Moharir, A.V. and Vijayraghavan, K.M., *J. Appl. Polym. Sci.*, 48, 1869, 1993.

249. Hebert, J.J. and Muller, L.L., *J. Appl. Polym. Sci.*, 18, 3373, 1974.

250. Rowland, S.P. and Bertoniere, N.R., Chemical methods of studying supramolecular structure, in *Cellulose Chemistry and Its Applications*, Nevell, T.P. and Zeronian, S.H., Eds., Ellis Horwood Ltd., Chichester, England, 1985, chap. 4, p. 112.

251. Bertoniere, N.R. and Zeronian, S.H., Chemical characterization of cellulose, in *The Structures of Cellulose*, Atalla, R.H., Ed., ACS Symposium Series, No. 340, American Chemical Society, Washington, D.C., 1987, p. 255.

252. Valentine, L., *Chem. Ind. (London)*, 47, 1279, 1956.

253. Jeffries, R., *J. Appl. Polym. Sci.*, 8, 1213, 1964.

254. Hailwood, A.J. and Horrobin, S., *Trans. Faraday Soc.*, 42B, 84, 1946.

255. Lewin, M., Guttman, H., and Saar, N., *Appl. Polym. Symp.*, 28, 791, 1976.

256. Lewin, M., New chemical approaches to the structure of cellulose in *Cellulose and Its Compounds*, Kennedy, J.F., Phillips, G.O., Wedlock, D.J., and Williams, P.A., Eds., Ellis Harwood, Ltd., Chichester, England, 1985, pp. 27–36.

257. Lewin, M. and Ben Bassat, A., SIRTEC, First International Symposium on Cotton Research, Paris, Institute Textiles de France, 1969, pp. 535–556.

258. Lewin, M., Guttmann, H., and Shabtai, D., *Appl. Polym. Symp.*, 31, 163, 1977.

259. Lewin, M., Guttmann, H., and Derfler, D., *J. Appl. Polym. Sci.*, 27, 3199, 1982.

260. Lewin, M., Guttmann, H., Knoll, A., and Derfler, D., *J. Polym. Sci., Polym. Chem. Ed.*, 20, 929, 1982.

261. Zeronian, S.H., Coole, M.L., Alger, K.W., and Chandler, J.M., *J. Appl. Polym. Sci., Appl. Polym. Symp.*, 37, 1053, 1983.

262. Nelson, M.L., *J. Polym. Sci.*, 43, 351, 1960.

263. Nickerson, R.F., *Text. Res. J.*, 21, 195, 1951.

264. Marchessault R.H. and Howsmon J.A., *Text. Res. J.*, 27, 30, 1957.

265. Rowland, S.P. and Pittman, P.F., *Text. Res. J.*, 35, 421, 1965.

266. Jeffries, R., Roberts, J.G., and Robinson, R.N., *Text. Res. J.*, 38, 234, 1968.

267. Cousins E.R., Bullock, A.L., Mack, C.H., and Rowland, S.P., *Text. Res. J.*, 34, 953, 1964.

268. Rowland, S.P. and Howley, P.S., *J. Polym. Sci., Poly. Sci. Ed.*, 23, 183, 1985.

269. Bertoniere, N.R., Howley, P.S., Ruppenicker, G.F., Anthony, W.S., and Hughs, S.F., Effect of ginning, greige mill, and wet processing on the microstructure of cotton fiber, Proceedings of the Beltwide Cotton Conference, 3, 1267–1269, 1992.

270. Guthrie, J.D., The chemistry of lint cotton, in *Chemistry and Chemical Technology of Cotton*, Ward, K., Ed., Interscience Publishers, New York, 1955, pp. 3–13.

271. Rollins, M.L., *Text. Res. J.*, 15, 65, 1945.

272. Flint, E.G., *Biol. Rev.*, 25, 414, 1950.

273. Warwicker, J.O., Simmens, S.C., and Hallam, P., *Text. Res. J.*, 40, 1051, 1970.

274. Goynes, W.R., Ingber, B.F., and Triplett, B.A., *Text. Res. J.*, 65, 4008, 1995.

275. Rollins, M.L., Moore, A.T., Goynes, W.R., Carra, J.H., and deGruy, I.V., *Am. Dyest. Rep.*, 54, 36, 1965.
276. Tripp, V.W., Moore, A.T., deGruy, I.V., and Rollins, M.L., *Text. Res. J.*, 30, 140, 1960.
277. Gipson, J.R. and Ray, L.L., *Cotton Grow. Rev.*, 47, 257, 1970.
278. Haigler, C.H., Rao, N.R., Roberts, E.M., Huang, J.-Y., Upchurch, D.R., and Trolinder, N.L., *Plant Physiol.*, 95, 88, 1991.
279. Kassenback, P., Incidence of Bilateral Structure of Cotton Fibers, in *First International Symposium on Cotton Textile Research*, Paris, 1969, p. 455.
280. Boylston, E.K. and Hebert, J.J., *Text. Res. J.*, 53, 469, 1983.
281. Welo, W.H., Ziffle, H.M., and McDonald, A.W., *Text. Res. J.*, 22, 261, 1952.
282. Jayme, G., *Tappi J.*, 41, 180A, 1958.
283. Bertoniere, N.R., *Modern Textile Characterization Methods*, Raheel, M., Ed., Marcel Dekker, New York, 1996, chap. 5, pp. 265–290.
284. Stone, J.E. and Scallan, A.M., *Cellul. Chem. Technol.*, 2, 343, 1968.
285. Martin, L.F. and Rowland, S.P., *J. Chromatogr.*, 28, 139, 1967.
286. Bredereck, K. and Blüher, A., *Melliand Textilberichte*, 73, E279 (English), 652 (German), 1992.
287. Ladisch, C.M., Yang, Y., Velayudhan, A., and Ladisch, M.R., *Text. Res. J.*, 62, 361, 1992.
288. Nelson, R. and Oliver, D.W., *J. Polym. Sci., Part C*, 36, 305, 1968.
289. Bertoniere, N.R., King, W.D., and Hughs, S.E., *Lignocellulosics—Science, Technology, Development and Use*, Kennedy, J.F., Phillips, G.O., and Williams, P.A., Eds., Ellis Horwood Ltd., Chichester, England, 1992, p. 457.
290. Bertoniere, N.R. and King, W.D., *Text. Res. J.*, 59, 114, 1989.
291. Blouin, F.A., Martin, L.F., and Rowland, S.P., *Text. Res. J.*, 40, 809, 1970.
292. Rowland, S.P., Wade, C.P., and Bertoniere, N.R., *J. Appl. Polym. Sci.*, 29, 3349, 1984.
293. Bertoniere, N.R., King, W.D., and Rowland, S.P., *J. Appl. Polym. Sci.*, 31, 2769, 1986.
294. Blouin, F.A., Martin, L.F., and Rowland, S.P., *Text. Res. J.*, 40, 959, 1970.
295. Bertoniere, N.R. and King, W.D., *Text. Res. J.*, 60, 606, 1990.
296. Bertoniere, N.R. and King, W.D., *Text. Res. J.*, 59, 608, 1989.
297. Bertoniere, N.R. and King, W.D., *Text. Res. J.*, 62, 349, 1992.
298. Bertoniere, N.R., King, W.D., and Welch, C.M., *Text. Res. J.*, 64, 247, 1994.
299. Ladisch, C.M. and Yang, Y., *Text. Res. J.*, 62, 481, 1992.
300. Rousselle, M.A., Bertoniere, N.R., Howley, P.S., and Goynes, W.R., Jr., *Text. Res. J.*, 72, 963, 2002.
301. Rousselle, M.A., Bertoniere, N.R., and Howley, P.S., *Text. Res. J.*, 73, 921, 2003.
302. Li, C., Ladisch, C.M., and Ladisch, M.R., *Text. Res. J.*, 71, 407, 2001.
303. Zeronian, S.H., Intercrystalline swelling of cellulose, in *Cellulose Chemistry and Its Applications*, Nevell, T.P. and Zeronian, S.H., Eds., Ellis Horwood Ltd., Chichester, England and Halsted Press, New York, 1985, chap. 5, pp. 138–158.
304. Jeffries, R., *J. Text. Inst.*, 51, T441, 1960.
305. Babbett, J.D., *Can. J. Res.*, A20, 143, 1943.
306. Zeronian, S.H. and Kim, M.S., Proceedings of the 1987 International Dissolving and Specialty Pulps Conference, TAPPI Press, Atlanta, GA, 1987, 125.
307. Weatherwax, R.C., *J. Colloid Interface Sci.*, 49, 40, 1974.
308. Mercer, J., British Patent 13296, 1850.
309. Abrahams, D.H., *Am. Dyestuff Reptr.*, 83(9), 78, 1994.
310. Freytag, R. and Donzé, J.-J., Alkali treatment of cellulose fibers, in *Handbook of Fiber Science and Technology*, Vol. 1, Part A, Lewin, M. and Sello, S.B., Eds., Marcel Dekker, New York, 1983, pp. 93–165.
311. Zeronian, S.H., Intracrystalline swelling of cellulose, in *Cellulose Chemistry and Its Applications*, Nevell, T.P. and Zeronian, S.H., Eds., Ellis Horwood Ltd., Chichester, England and Halsted Press, New York, 1985, chap. 6, pp. 159–180.
312. Vigo, T.L., *Textile Processing and Properties*, Elsevier, Amsterdam, 1994.
313. Zeronian, S.H. and Cabradilla, K.E., *J. Appl. Polym. Sci.*, 17, 539, 1973.
314. Aboul-Fadl, S.M., Zeronian, S.H., Kamal, M.M., Kim, M.S., and Ellison, M.S., *Text. Res. J.*, 55, 461, 1985.

315. Zeronian, S.H., *J. Appl. Polym. Sci., Appl. Polym. Symp.*, 47, 445, 1991.
316. Sarko, A., Recent x-ray crystallographic studies of celluloses, in *Cellulose: Structure, Modification and Hydrolysis*, Young, R.A. and Rowell R.M., Eds., John Wiley & Sons, New York, 1986, pp. 29–49.
317. Stevens, C.V. and Roldán-González, L.G., Liquid ammonia treatment of textiles, in *Handbook of Fiber Science and Technology*, Vol. 1, Part A, Lewin, M. and Sello, S.B., Eds., Marcel Dekker, New York, 1983, pp. 167–203.
318. Bredereck, K., *Mell. Textilber.*, 72(6), 446, 1991.
319. Lewin, M. and Roldan, L.G., *J. Polymer Sci., Part C*, 36, 213, 1971.
320. Lewin, M., Rau, R.O., and Sello, S.B., *Text. Res. J.*, 44, 680, 1974.
321. Schuerch, C., *For. Prod. J.*, 14, 377, 1964.
322. Schuerch, C., *Ind. Eng. Chem.*, 55, 139, 1963.
323. Schuerch, C., U.S. Patent 3282313, 1966.
324. Schuerch, C., Budrick, M.P., and Mahdalik, M., *I & E.C. Prod. Res. Dev.*, 5, 101, 1966.
325. Pentoney, R.C., *I & E.C. Prod. Res. Dev.*, 5, 105, 1966.
326. Norwegian Patent 152995, 1966.
327. Coats, J. and Coats, P., British Patent 1136417, 1967.
328. Majewicz, T.G. and Podlas, T.J., Cellulose ethers, in *Encyclopedia of Chemical Technology*, 4th ed., Vol. 5, John Wiley & Sons, New York, 1993, pp. 541–563.
329. Savage, A.B., Klug, E.D., Bikales, N.M., and Stanonis, D.J., Cellulose ethers, in *Encyclopedia of Polymer Science and Technology*, Vol. 3, Interscience, New York, 1966, pp. 459–549.
330. Nicholson, M.D. and Merritt, F.M., Cellulose ethers, in *Cellulose Chemistry and Its Applications*, Ellis Horwood Ltd., Chichester, England, 1985, chap. 15, pp. 363–383.
331. Vail, S.L., Crosslinking of cellulose, in *Cellulose Chemistry and Its Applications*, Ellis Horwood Ltd., Chichester, England, 1985, chap. 16, pp. 384–422.
332. Bikales, N.M., Cyanoethylation, in *Encyclopedia of Polymer Science and Technology*, Vol. 4, Interscience, New York, 1966, pp. 533–562.
333. Petersen, H., Crosslinking chemicals and the chemical principles of the resin finishing of cotton, in *Chemical Aftertreatment of Textiles*, Mark, H., Wooding, N.S., and Atlas, S.M., Eds., Wiley Interscience, New York, 1971, pp. 135–233.
334. Ryan, J.J., Wash-and-wear fabrics, in *Chemical Aftertreatment of Textiles*, Mark, H., Wooding, N.S., and Atlas, S.M., Eds., Wiley Interscience, New York, 1971, pp. 417–464.
335. Heap, S.A., Hunt, R.E., Rennison, P.A., and Tattersall, R., Polycondensation products: urea-formaldehyde resin in the treatment of textiles, in *Chemical Aftertreatment of Textiles*, Mark, H., Wooding, N.S., Atlas, S.M., Eds., Wiley Interscience, New York, 1971, pp. 267–317.
336. Herbes, W.F., O'Brien, S.J., and Weyker, R.G, Polycondensation products: melamine-formaldehyde, in *Chemical Aftertreatment of Textiles*, Mark, H., Wooding, N.S., and Atlas, S.M., Eds., Wiley Interscience, New York, 1971, pp. 319–329.
337. Weyker, R.G., O'Brien, S.J., and Herbes, W.F., Polycondensation products: methylol compounds, in *Chemical Aftertreatment of Textiles*, Mark, H., Wooding, N.S., and Atlas, S.M., Eds., Wiley Interscience, New York, 1971, pp. 331–355.
338. Dolby, P.J., *Text. Chem. Color.*, 9(11), 32, 1977; Hilderbrand, D.R., *CHEMTECH*, 8, 224, 1978.
339. Aspland, J.R., Reactive dyes and their application, *Text. Chem. Color.*, 24(5), 31–36, 1992; Aspland, J.R., Practical application of reactive dyes, *Text. Chem. Color.*, 24(6), 35–40, 1992.
340. Reactive dyes, in *Cotton Dyeing and Finishing: A Technical Guide*, Cotton Incorporated, Cary, NC, 1996, pp. 76–92.
341. Shore, J., Dyeing with reactive dyes, in *Cellulose Dyeing*, Shore, J., Ed., The Society of Dyers and Colourists, Bradford, UK, 1995, chap. 4.
342. Siegel, E., Reactive dyes: reactive groups, in *The Chemistry of Synthetic Dyes*, Vol. 6, Venkataraman, K., Ed., Academic Press, New York, 1972, pp. 182–194.
343. Horrocks, A.R., *J. Soc. Dyers Colour*, 16, 62, 1986.
344. Lewin, M. and Sello, S.B., *Flame Retardant Polymeric Materials*, Vol. 1, Lewin, M., Atlas, S.M., and Pierce, E.M., Eds., Plenum Press, New York, 1975, pp. 19–136.

345. Lewin, M. and Bash, A., *Flame Retardant Polymeric Materials*, Vol. 2, Lewin, M., Atlas, S.M., and Pearce, A.M., Eds., Plenum Press, New York, 1978, pp. 1–42.
346. Lewin, M., *Handbook of Fiber Science and Technology*, Vol. 2, *Chemical Processing of Fibers and Fabrics, Part B: Functional Finishes*, Lewin, M. and Sello, S.B., Eds., Marcel Dekker, New York, 1984, pp. 1–141.
347. Wakelyn, P.J., Adair, P.K., and Barker, R.H., *Fire Mater.*, 29, 15, 2005.
348. Miller, B., Martin, J.R., and Turner, R., *J. Appl. Polym. Sci.*, 28, 45, 1983.
349. Ohlemiller, T.J., *Combust. Sci. Technol.*, 26, 89, 1981.
350. Miller, B., Martin, J.R., and Meiser, C.H., Jr., *J. Appl. Polym. Sci.*, 17, 629, 1973.
351. McCarter, R.J., *J. Consumer Prod. Flammability*, 4, 346, 1977.
352. Krasney, J.F., *Text. Chemist and Colorist*, 24(11), 12, 1992.
353. Hshieh, F.Y. and Richards, G.N., *Combust. Flame*, 80, 395, 1989.
354. Hshieh, F.Y. and Richards, G.N., *Combust. Flame*, 76, 49, 1989.
355. Shafizadeh, F. and Sekiguchi, Y., *Combust. Flame*, 55, 171, 1984.
356. Koenig, P.A., Neumeyer, J.P., Knoepfler, N.B., and Vix, H.L.E., Monograph, Organic Coatings and Plastics Chemistry, Vol. 33, No. 1, Presented at the 165th American Chemical Society Meeting, April 1973, pp. 476–483.
357. Weast, R.C., Ed., *Handbook of Chemistry and Physics*, 46th ed., Chemical Rubber Company., Cleveland, OH, 1965, p. C-247.
358. Wakelyn, P.J., Rearick, W.A., and Turner, J., *Am. Dyestuff Reptr.*, 87(2), 13, 1998.
359. Hooper, N.K. and Ames, B.N., *Regulation of Cancer-Causing Flame-Retardant Chemicals and Governmental Coordination of Testing of Toxic Chemicals*, Serial No. 95–33, U.S. Government Printing Office, Washington, D.C., 1977, 42; Studies show flame retardants break down, data said to refute previous industry studies, *BNA Daily Report for Executives*, November 24, 2003, p. 24.
360. Calamari, T.A., Jr. and Harper, R.J., Jr., Flame retardants for textiles, in *Kirk–Othmer Encyclopedia of Chemical Technology*, 4th ed., Vol. 10, Kroschivitz, J.I. and Howe-Grant, M., Eds., John Wiley & Sons, New York, 1993, p. 998.
361. Weil, E.D., *Flame Retardant Polymeric Materials*, Vol. 2, Lewin, M., Atlas, S.M., and Pearce, E.M., Eds., Plenum Press, New York, 1978, p. 123.
362. Beninate, J.V., Boylston, E.K., Drake, G.L., and Reeves, W.A., *Am. Dyest. Rep.*, 57(25), 981, 1968.
363. A performance comparison—Indura Proban cotton vs. Nomex, Westex, Inc., Chicago, IL, 1991.
364. Tesoro, G.C. and Meiser, C., Jr., *Text. Res. J.*, 40, 430, 1970.
365. Kruse, W., *Melliand Textilber*, 50, 460, 1969.
366. Baer, E., Proceedings of the Symposium on Textile Flammability, Lelanc Research Corp., East Greenwich, RI, 1973, p. 117.
367. Mischutin, V., Proceedings of the Symposium on Textile Flammability, LeBlanc, R.B., Ed., LeBlanc Research Corp., East Greenwich, RI, 1975, p. 211.
368. Harper, R.J., Jr., Ruppenicker, G.F., Jr., and Donaldson, D.J., *Text. Res. J.*, 56, 80, 1986.
369. Ruppenicker, G.F., Jr., Harper, R.J., Jr., Sawhney, A.P.S., and Robert, K.Q., *Text. Technol. Forum*, 90, 71, 1990.
370. Sawhney, A.P.S., Ruppenicker, G.F., and Price, J.B., *Text. Res. J.*, 68, 203, 1998.
371. Sawhney, A.P.S., Ruppenicker, G.F., Calamari, T.A., and Parachuru, R., A fire barrier of predominately-cotton content, Proceedings of the Beltwide Cotton Conference, 1365, 1999.
372. Ruppenicker, G.F. and Sawhney, A.P.S., *Text. Sci.*, 93, 642, 2001.
373. Berkley, R., *Text. Chem. Color.*, 25(5), 18, 1993.
374. Donaldson, D.J. and Harper, R.J., Jr., *J. Consum. Prod. Flam.*, 7(1), 40, 1980.
375. Wakelyn, P.J., Adair, P.K., and Wolf, S., Cotton and cotton modacrylic blended batting fire-blocking barriers for soft furnishings to meet federal and state flammability standards, Proceedings of the 2004 Beltwide Cotton Conferences, National Cotton Council, Memphis, TN, 2004, pp. 2829–2842.
376. Welch, C.M. and Andrews, B.K., U.S. Patent 4820307, 1989.
377. Blanchard E.J., Graves, E.E., and Salame, P.A., *J. Fire Sci.*, 18, 151, 2000.
378. Blanchard, E.J. and Grawes, E.E., *Colourage Annu.*, 49, 17, 2001.
379. Blanchard, E.J. and Grawes, E.E., *Text. Res. J.*, 72, 39, 2002.

380. Rearick, W.A., Wallace, J.L., Martin, V.B., and Wakelyn, P.J., *AATCC Rev.*, 2, 12, 2002.
381. Vigo, T.L. and Bruno, J.S., Applications of fibrous substrates containing insolubilized polymers, Proceedings Technology 2002: Third NASA Natl. Tech. Trans. Conf., 2, 307, 1992.
382. Vigo, T.L. and Bruno, J.S., Fibers with multifunctional properties: a holistic approach, in *Handbook of Fibers and Science Technology*, Vol. III, High Technology Fibers, Part C, Lewin, M. and Preston, J., Eds., 1993, pp. 3–356.
383. Gagliardi, D.D. and Shippee, F.B., *Am. Dyestuff Reptr.*, 52, 300, 1963.
384. Rowland, S.P., Welch, C.M., Brannan, M.A.F., and Gallagher, D.M., *Text. Res. J.*, 37, 933, 1967.
385. Welch, C.M., *Text. Res. J.*, 58, 480, 1988.
386. Welch, C.M. and Andrews, B.A.K., U.S. Patent 4936865, 1990.
387. Welch, C.M. and Andrews, B.A.K., U.S. Patent 4975209, 1990.
388. Welch, C.M. and Andrews, B.A., *Text. Chem. Color.*, 21(2), 13, 1989.
389. Welch, C.M., *Text. Chem. Color.*, 22(5), 13, 1990.
390. Welch, C.M., *Surface Characteristics of Fibers and Textiles*, Pastore, D.M. and Kiekens, P., Eds., Marcel Dekker, New York, 2000, chap. 1, pp. 1–32.
391. Isaacs, P., Lewin, M., Sello, S.B., and Stevens, C.V., *Text. Res. J.*, 44, 700, 1974.
392. Lewin, M., Isaacs, P., Sello, S.B., and Stevens, C., *Textilveredlung*, 8, 158, 1973.
393. Lewin, M. and Isaacs, P., German Patent 2127188, 1972.
394. Buras, E.M., Cooper, A.S., Keating, E.J., and Goldthwait, C.F., *Am. Dyestuff Reptr.*, 43(7), P203, 1954.
395. Buras, E.M., Hobart, S.R., Hamalainan, C., and Cooper, A.S., *Text. Res. J.*, 27, 214, 1957.
396. Andrews, B.A.K., *Text. Chem. Color.*, 22(9), 63, 1990.
397. Welch, C.M., *Am. Dyestuff Reptr.*, 83(9), 19, 1994.
398. Yang, C.Q., *Text. Res. J.*, 61, 433, 1991.
399. LeBlanc, R.B. and LeBlanc, D.A., *Text. Chem. Color.*, 6(10), 29, 1974.
400. Little, R.W., Ed., *Flameproofing Textile Fabrics*, Series, No. 104, ACS Monograph, Reinhold, New York, 1947, pp. 196 and 198.
401. LeBlanc, R.B. and Symm, R.H., U.S. Patent 3409463, to Dow Chemical Company, 1968.
402. Edward, J.V., Batiste, S.L., Gibbins, E.M., and Goheen, S.C., *J. Peptide Res.*, 54, 536, 1999.
403. Edwards, J.V., Yager, D.R., Cohen, I.K., Diegelmann, R.F., Montante, S., Bertoniere, N., and Bopp, A.F., *Wound Rep. Reg.*, 9, 50, 2001.
404. Edwards, J.V., Cohen, I.K., Diegelmann, R.F., and Yager, D., Wound Dressings with Protease-Lowering Activity, U.S. Patent 6627785, 2003.
405. Edwards, J.V., Sethumadhavan, K., and Ullah, A.H.J., *Bioconjugate Chem.*, 11, 469, 2000.
406. Grimsley, J.K, Singh, W.P., Wild, J.R., and Giletto, A., A novel enzyme–based method for the wound–surface removal and decontamination of organophosphorus nerve agents, in *Bioactive Fibers and Polymers*, Edwards, J.V. and Vigo, T.L., Eds., ACS Symposium Series, No. 792, American Chemical Society, Washington D.C., 2001, pp. 35–49.
407. Edwards, J.V., Eggleston, G., Yager, D.R., Cohen, I.K., Diegelmann, R.F., and Bopp, A.F., *Carbohydr. Polym.*, 50, 305, 2002.
408. Edwards, J.V., Bopp, A.F., Batiste, S.L., and Goynes, W.R., *J. Biomed. Mater. Res.*, 66A, 433, 2003.
409. Slater, K., *Text. Prog.*, 23/1/2/3, 1, 1991.
410. Whistler, R.L. and BeMiller, J.N., Eds., *Methods in Carbohydrate Chemistry*, Academic Press, New York, Vol. III, 1963, p. 31274; Vol. V, 1965, p. 249; Vol. VI, 1972, p. 76.
411. Girard, A., *Compt. Rend.*, 81, 1105, 1875.
412. Witz, G., *Bull. Soc. Ind. Rouen*, 10, 416, 1982; 11, 169, 1982.
413. Cross, C.F. and Bevan, E.J., *J. Soc. Chem. Ind.*, 3, 206, 291, 1884.
414. Shafizadeh, F., *Cellulose Chemistry and Its Applications*, Nevell, T.P. and Zeronian, S.H., Eds., Ellis Horwood Ltd., Chichester, England, 1985, chap. 11, pp. 266–289.
415. Jayme, G. and Verburg, W., *Reyon, Zellwolle, Chemiefasern*, 32, 193, 275, 1951.
416. Isbell, H.S., *J. Res. NBS*, 32, 54, 1944; *Ann. Rev. Biochem.*, 12, 205, 1943.
417. Davidson, G.F., *J. Text. Inst.*, 43, T291, 1952; *J. Text. Inst.*, 29, T195, 1938.
418. Lewin, M. and Epstein, J.A., *J. Polym. Sci.*, 58, 1023, 1962.

419. Lewin, M. and Albeck, M., *Mild Oxidation of Cotton*, Final Report FG-IS-101–58, Submitted to ARS, USDA, Jerusalem, 1963.
420. Lewin, M., *Handbook of Fiber Science and Technology*, Vol. 1, *Chemical Processing of Fibers and Fabrics, Fundamentals and Preparation*, Part B, Lewin, M. and Sello, S.B., Eds., Marcel Dekker, New York, 1984.
421. Mitchel, R.L., *Ind. Eng. Chem.*, 38, 843, 1946.
422. Cumberbirch, R.J.E. and Harland, W.G., *Shirley Inst. Mem.*, 31, 199, 1958.
423. Samuelson, O., *Methods in Carbohydrate Chemistry*, Whistler, R.L. and BeMiller, J.N., Eds., Academic Press, New York, 1963, p. 31.
424. Samuelson, O. and Wennerblom, A., *Svensk. Papperstidn.*, 58, 713, 1955.
425. Sobue, H. and Okubo, M., *Tappi J.*, 39, 415, 1956.
426. Davidson, G.F., *J. Text. Inst.*, 39, T65, 1948.
427. Ludtke, M., *Angew. Chem.*, 48, 650, 1935.
428. Nabar, G.M. and Padmanabhan, C.V., *Proc. Indian Acad. Sci.*, 31A, 371, 1950.
429. Lefevre, K.V. and Tollens, B., *Ber. Dtsch. Chem. Ges.*, 40, 4513, 1907.
430. Andersen, D.M.W., *Talanta*, 2, 73, 1959.
431. Nevell, T.P., *Methods in Carbohydrate Chemistry*, Vol. III, Whistler, R.L. and BeMiller, J.N., Eds., Academic Press, New York, 1963, p. 49.
432. Swalbe, C.G., *Ber. Dtsch. Chem. Ges.*, 40, 1347, 1907.
433. Clibbens, D.A., *J. Text. Inst.*, 45, P173, 1954.
434. Colbran, R.L. and Davidson, G.F., *J. Text. Inst.*, 52, T291, 1961.
435. Davidson, G.F. and Nevell, T.P., *J. Text. Inst.*, 46, T407, 1955.
436. Davidson, G.F. and Nevell, T.P., *J. Text. Inst.*, 48, T356, 1957.
437. Ellington, A.C. and Purves, C.B., *Can. J. Chem.*, 31, 801, 1953.
438. Lewin, M., *Methods in Carbohydrate Chemistry*, Vol. VI, Whistler, R.L. and BeMiller, J.N., Eds., Academic Press, New York, 1972, p. 76.
439. Schmorak, J. and Lewin, M., *Anal. Chem.*, 33, 1403, 1961.
440. Godsay, M. and Lewin, M., Cellulose and wood chemistry and technology; Proceedings of the 10th Cellulose Conference, Schuerch, C., Ed., Wiley Interscience, Syracuse, 1990, pp. 1059–1084.
441. Albeck, M., Ben-Bassat, A., and Lewin, M., *Text. Res. J.*, 35, 935, 1965.
442. Lewin, M., *Mild Oxidation of Cotton*, Project FG-IS-109, Final Report, submitted to the ARS USDA, Jerusalem, 1968.
443. Lewin, M. and Ettinger, A., *Cellul. Chem. Technol.*, 3, 9, 1969.
444. Isbell, H.S., *Methods in Carbohydrate Chemistry*, Vol. V, Whistler, R.L. and BeMiller, J.N., Eds., Academic Press, New York, 1965, p. 249.
445. BeMiller, J.N., *Adv. Carbohydr. Chem.*, 22, 25, 1967.
446. Sharples, A., *J. Polym. Sci.*, 14, 95, 1954.
447. Daruwalla, E.H. and Narsion, M.G., *Tappi J.*, 49, 106, 1966.
448. Michie, R.I., Sharples, A., and Walter, A.A., *J. Polym. Sci.*, 51, 85, 1961.
449. Sippel, A., *Das Papier*, 13, 413, 1959.
450. Nelson, M. and Tripp, V.W., *J. Polym. Sci.*, 10, 577, 1953.
451. Batista, O.A., *Microcrystal Polymer Science*, McGraw-Hill, New York, 1975.
452. Wood, B.F., Conner, A.H., and Hill, C.G., Jr., *J. Appl. Polym. Sci.*, 37, 1373, 1989.
453. Lin, C.H., Conner, A.H., and Hill, C.G., Jr., *J. Appl. Polym. Sci.*, 42, 417, 1991.
454. Nevell, T.P. and Upton, W.R., *Carbohydr. Res.*, 49, 163, 1976.
455. Davidson, G.F., *J. Text. Inst.*, 25, T174, 1934.
456. Richards, G.N. and Shepton, H.H., *J. Chem. Soc.*, 64, 4492, 1957.
457. Machell, G.N. and Richards, G.N., *J. Chem. Soc.*, 64, 4500, 1957.
458. Machell, G.N. and Richards, G.N., *J. Chem. Soc.*, 69, 1932, 1960.
459. Lobry de Bruyn, C.A. and Alberda van Ekenstein, W., *Recl. Trav. Chim.*, 14, 201, 1895; 16, 262, 1897; *Ber. Dtsch. Chem. Ges.*, 28, 3078, 1895.
460. Speck, J.C., *Adv. Carbohydr. Chem.*, 13, 63, 1958.
461. Corbett, W.M., *J. Soc. Dyers Col.*, 76, 265, 1960.
462. Haas, D.W., Hrutfiord, B.F., and Sarkanen, K.V., *J. Appl. Polym. Sci.*, 11, 587, 1967.

463. Lai, Y.-Z. and Outto, D.E., *J. Appl. Polym. Sci.*, 23, 3219, 1979.

464. Johansson, M.H. and Samuelson, O., *J. Appl. Polym. Sci.*, 22, 615, 1978.

465. Albeck, A., Ben-Bassat, A., Epstein, J.A., and Lewin, M., *Text. Res. J.*, 35, 836, 1965; *Isr. J. Chem.*, 1(3a), 304, 1963.

466. Lewin, M., *Text. Res. J.*, 35, 979, 1965.

467. Ziderman, I., Bel-Aiche, J., Basch, A., and Lewin, M., *Carbohydr. Res.*, 43, 255, 1975.

468. Lewin, M., Ziderman, I., Weiss, N., Basch, A., and Ettinger, A., *Carbohydr. Res.*, 62, 393, 1978.

469. Ziderman, I., *Cellul. Chem. Technol.*, 14, 703, 1980.

470. Lewin, M. and Ziderman, I., *Cellul. Chem. Technol.*, 14, 743, 1980.

471. Gascoigne, J.A. and Gascoigne, M.M., *Biological Degradation of Cellulose*, Butterworth & Co., London, England, 1960.

472. Finch, P. and Roberts, J.C., Enzymatic degradation of cellulose, in *Cellulose Chemistry and Its Applications*, Nevell, T.P. and Zeronian, S.H., Eds., Ellis Horwood Ltd., Chichester, England, 1985, chap. 13, pp. 312–343.

473. Griffin, D.H., *Fungal Physiology*, Wiley–Liss, New York, 1994, pp. 160–163.

474. Misaghi, I.J., *Physiology and Biochemistry of Plant–Pathogen Interactions*, Plenum Press, New York, 1982, p. 25.

475. Leschine, S.B., *Annu. Rev. Microbiol.*, 49, 399, 1995.

476. Weimer, P.J. and Odt, C.L., Cellulose degradation by ruminal microbes: physiological and hydrolytic diversity among ruminal cellulolytic bacteria, in *Enzymatic Degradation of Insoluble Carbohydrates*, Saddler, J.N. and Penner, M.H., Eds., ACS Symposium Series, No. 618, American Chemical Society, Washington, D.C., 1995, chap. 18.

477. Stanier, R.Y., Adelberg, E.A., and Ingraham, J.L., *The Microbial World*, Prentice Hall, Englewood Cliffs, NJ, 1976, pp. 780–781.

478. Hamby, D.S., Ed., *The American Cotton Handbook*, 3rd ed., Vol. I, Interscience, New York, 1965, chap. 2, p. 35.

479. Hall, L.T. and Elting, J.P., *Text. Ind.*, 117, 100, 1953.

480. Hegn, A.N.J., *Text. Res. J.*, 120, 137, 1956.

481. Marsh, P.B., Guthrie, L.R., and Butler, M.L., *Text. Res. J.*, 21, 565, 1951.

482. Hegn, A.N.J., *Text. Res. J.*, 27, 591, 1957.

483. Chun, D., Use of high cotton moisture content during storage to reduce stickiness, Proceedings of the Beltwide Cotton Research Conference, Vol. 2, 1997, pp. 1642–1648.

484. Allen, S.J., Auer, P.D., and Pailthorpe, M.T., *Text. Res. J.*, 65, 379, 1995.

485. Interpretation and substantiation of environmental claims, U.S. Code of Federal Regulations, 16, Part 260, p. 5.

486. U.S. EPA, Federal Register, Vol. 57, November 19, 1992, p. 54456.

487. *1988 Book of ASTM Standards*, American Society for Testing and Materials, Philadelphia, PA, 1988.

488. Moreau, J.P., *J. Nonwoven Res.*, 10, 14–22, 1990; Goynes, W.R., Delucca, A.J., and Moreau, J.P., Structural evaluation of biodegradability of nonwoven fabrics, Proceedings of the 49th Annual Meeting of Electron Microscopy Society of America, 1991.

489. OECD guidelines for testing of chemicals, Organization for Economic Co-operation and Development, Paris, France, 1992.

490. Shafizadeh, F., *Advances in Carbohydrate Chemistry* Vol. 23, Wolfrom, M.C. and Tipson, R.S., Eds., Academic Press, New York, 1968, p. 419.

491. Madorsky, S.L., Hart, V.E., and Strauss, S., *J. Res. Natl. Bur. Stand.*, 56, 343, 1956; 58, 343, 1958.

492. Lipska, A.E. and Parker, W.J., *J. Appl. Polym. Sci.*, 10, 1439, 1965.

493. Akita, K., *Rep. Fire Res. Inst. Jpn.*, 9, 10, 1959.

494. Stamm, A.J., *Ind. Eng. Chem.*, 48, 413, 1956.

495. Tang, W.K. and Neil, W., *J. Polym. Sci.*, C, 6, 65, 1964.

496. Chatterjee, P.K. and Conrad, C.M., *Text. Res. J.*, 36, 487, 1966.

497. Chatterjee, P.K., *J. Appl. Polym. Sci.*, 12, 1859, 1968.

498. Basch, A. and Lewin, M., *J. Polym. Sci., Polym. Chem. Ed.*, 11, 3077–3093, 3095–3107, 1973; 12, 2053–2063, 1974; 13, 493–499, 1975.

499. Lewin, M., Basch, A., and Roderig, C., Proceedings of the International Symposium on Macromolecules, Rio de Janeiro, Mano, E., Ed., Elsevier, Amsterdam, 1975, pp. 225–250.
500. Lewin, M. and Basch, A., *Encyclopedia of Polymer Science and Technology*, Suppl. 2, Wiley, New York, 1977, pp. 340–363.
501. Tyler, D.N. and Wooding, N.S., *J. Soc. Dyers Col.*, 74, 283, 1958.
502. Franklin, W.E. and Rowland, S.P., *J. Macromol. Sci., Chem.*, A19, 165, 1983.
503. Franklin, W.E., *J. Macromol. Sci., Chem.*, A19, 619, 1983.
504. Franklin, W.E., *J. Macromol. Sci., Chem.*, A21, 377, 1984.
505. Franklin, W.E., Proceedings of the Eleventh North American Thermal Analysis Society Conference, Vol. 11, 1981, p. 471.
506. Rushnak, I. and Tanczos, I., Preprints of papers presented at IUPAC Symposium on Macromolecules, Helsinki, Vol. 5, 1972, p. 127.
507. Back, F.L. and Klinga, L.O., *Svensk Pepperstidn*, 66, 745, 1965.
508. Back, F.L. and Didriksen, E.I., *Svensk Pepperstidn*, 72, 687, 1969.
509. Brushwood, D.E., *Text. Res. J.*, 58, 309, 1988.
510. Manley-Harris, M. and Richards, G.N., *Carbohydr. Res.*, 254, 195, 1994.
511. Phillips, G.O. and Arthur, J.C., Jr., Photochemistry and radiation chemistry of cellulose, in *Cellulose Chemistry and Its Applications*, Nevell, T.P. and Zeronian, S.H., Eds., Ellis Horwood Ltd., Chichester, England, 1985, chap. 12, pp. 290–311.
512. Baugh, P.J. and Phillips, G.O., *Cellulose and Cellulose Derivatives*, High Polymers, Volume 5, Part V, Bikales, N.M. and Segal, L., Eds., John Wiley & Sons, New York, 5, 1971, p. 1047.
513. McKellar, J.F., *Radiat. Res. Rev.*, Elsevier, Amsterdam, 3, 141, 1971.
514. Phillips, G.O., *The Carbohydrates*, 2nd ed., Vol. IB, Pigman, W. and Horton, D., Eds., Academic Press, New York, 1980, pp. 1217–1299.
515. Phillips, G.O., Baugh, P.J., McKellar, J.F., and Von Sonntag, C., Interaction of radiation with cellulose in the solid state, in *Cellulose Chemistry and Technology*, Arthur, J.C., Jr., Ed., ACS Symposium Series, No. 48, American Chemical Society, Washington, D.C., 1977, p. 313.
516. Launer H.F. and Wilson, W.K., *J. Am. Chem. Soc.*, 71, 958, 1949.
517. Egerton, G. S., *J. Soc. Dyers Colour*, 65, 764, 1949.
518. Appleby, D.K., *Am. Dyestuff Reptr.*, 38, 149, 1949.
519. Egerton, G.S., *Text. Res. J.*, 18, 659, 1948.
520. Bernard, W.N., Gremillion, S.G., Jr., and Goldthwait, C.F., *Text. Res. J.*, 26, 81, 1956.
521. Blouin, F.A. and Arthur, J.C., Jr., *J. Chem. Eng. Data*, 5, 470, 1960.
522. Arthur, J.C., Jr., Free-radical initiated graft polymerization of vinyl monomers onto cellulose, in *Graft Copolymerization of Lignocellulosic Fibers*, Hon, D.N.-S., Ed., ACS Symposium Series, No. 187, American Chemical Society, Washington D.C., 1982, p. 21.
523. Arthur, J.C., Graft polymerization onto polysaccharides, in *Advances in Macromolecular Chemistry*, Pasik, W.M., Ed., Academic Press, London, England, 1970, p. 1.
524. Arthur, J.C., Jr., Special properties of block and graft copolymers and applications in fiber form, in *Block and Graft Copolymers*, Burke, J.J. and Weiss, V., Eds., Syracuse University Press, Syracuse, New York, 1973, p. 295.
525. Blumenthal, W.B., *Ind. Eng. Chem.*, 42, 640, 1950.
526. Conner, C.J., Brysson, R.J., Walker, A.M., Harper, R.J., Jr., and Reeves, W.A., *Text. Chem. Color.*, 10(4), 17, 1978.
527. Matthews, J.M., *Application of Dyestuffs*, John Wiley & Sons, New York, 1920, p. 521.
528. Conner, C.J., Danna, G.S., Cooper, A.S., Jr., and Reeves, W.A., *Text. Res. J.*, 37(2), 94, 1967.
529. Kirk, O., Damhus, T., Borchert, T.V., Fuglsang, C.C., Olsen, H.S., Hansen, T.T., Lund, H., Schiff, H.E., and Nielsen, L.K., Enzyme applications (industrial), in *Kirk–Othmer Encyclopedia of Chemical Technology*, 5th ed., Vol. 9, John Wiley & Sons, New York, 2004.
530. Betrabet, S.M., *Colourage*, 41(5), 21, 1994.
531. Dominguez, R., Souchon, H., Spinelli, S., Dauter, Z., Wilson, K.S., Chauvaux, S., Begium, P., and Alizari, P.M., *Nat. Struct. Biol.*, 2, 569, 1995.
532. Rouvinen, J., Bergfors, T., Teeri, T., Knowles, J.K.C., and Jones, T.A., *Science*, 249, 380, 1990.
533. Davies, G. and Henrissat, B., *Structure*, 3, 853, 1995.

534. Buschle-Diller, G., Zeronian, S.H., Pan, N., and Soon, M.Y., *Text. Res. J.*, 64(56), 270, 1994.

535. Reese, E.T., Segal, L., and Tripp, V.W., *Text. Res. J.*, 27(8), 626, 1957.

536. Gascoigne, A. and Gascoigne, M.M., *Biological Degradation of Cellulose*, Butterworth & Co., London, Great Britain, 1960, p. 72.

537. Tyndall, M., *Text. Chem. Color.*, 24(6), 23, 1992.

538. Ajgaonkar, S.B., *Colourage*, 42(1), 35, 1995.

539. Cox, T., Hawks, P.E., and Klahorst, S.A., U.S. Patent 5232851, August 3, 1993.

540. Koo, H., Ueda, U., Wakida, T., Yoshimura, Y., and Igarashi, T., *Text. Res. J.*, 64(2), 70, 1994.

541. Thorsen, W.J., *Text. Res. J.*, 41, 331, 1971.

542. Thorsen, W.J., *Text. Res. J.* 41, 455, 1971.

543. Thorsen, W.J., *Text. Res. J.*, 44, 422, 1974.

544. Thibodeaux, D.P. and Copeland, H.R., Corona treating cotton: its relationship to processing performance and quality, *Proceedings of the 15th Textile Chemistry and Processing Conference*, New Orleans, 1975, 36.

545. Belin, R.E., *J. Text. Inst.*, 67, 249, 1976.

546. Abbott, G.M., *Text. Res. J.*, 47, 141, 1977.

547. Abbott, G.M. and Robinson, G.A., *Text. Res. J.*, 47, 199, 1977.

548. Anon., Preparation of cotton yarns, knits, and woven fabrics, *Cotton Dyeing and Finishing: A Technical Guide*, Cotton Incorporated, Cotton Incorporated, Cary, NC, 1996, pp. 1–26.

549. Rivlin, J., *The Dyeing of Textile Fibers—Theory and Practice*, Philadelphia College of Textiles and Science, Philadelphia, PA., 1992, 72.

550. Aspland, J.R., *Text. Chem. Color.*, 23(11), 41, 1991.

551. Aspland, J.R., *Text. Chem. Color.*, 24(3), 21, 1992.

552. Aspland, J.R., *Text. Chem. Color.*, 24(1), 22, 1992.

553. Horne, C.M., *Text. Chem. Color.*, 27(12), 27, 1995.

554. Aspland, J.R., *Text. Chem. Color.*, 24(8), 26, 1992; Aspland, J.R., *Text. Chem. Color.*, 24(9), 74, 1992.

555. Gregory, P., Dyes and dye intermediates, in *Kirk–Othmer Encyclopedia of Chemical Technology*, 4th ed., Vol. 8, John Wiley & Sons, New York, 1993, p. 542.

556. McGregor, R., *Text. Chem. Color.*, 12(12), 19, 1980.

557. Anon., Dyes, in *Cotton Dyeing and Finishing: A Technical Guide*, Cotton Incorporated, Cary, NC, 1996. pp. 28–132.

558. Bide, M., Dyes, in *Kirk–Othmer Encyclopedia of Chemical Technology*, 5th ed., Vol. 8, John Wiley & Sons, New York, 2004.

559. Dolby, P.J., *Text. Chem. Color.*, 9(11), 264, 1977.

560. Aspland, J.R., *Text. Chem. Color.*, 25(10), 31, 1993.

561. Blanchard, E.J. and Reinhardt, R.M., *Text. Chem. Color.*, 24(1), 13, 1992.

562. Harper, R.J., Jr., Blanchard, E.J., Allen, H.A., Reinhardt, R.M., Cheek, L., English, S., Etters, J.N., Hsu, L.H., and Roussel, L., *Text. Chem. Color.*, 20(5), 25, 1988.

563. Reinhardt, R.M. and Blanchard, E.J., *Am. Dyestuff Reptr.*, 79(6), 15, 1990.

564. Blanchard, E.J., Reinhardt, R.M., Andrews, B.A.K, *Text. Chem. Color.*, 23(5), 25, 1991.

565. Andrews, B.A.K., Blanchard, E.J., and Reinhardt, R.M., *Am. Dyestuff Reptr.*, 79(9), 48, 1990.

566. Lord, E. and Heap, S.A., *The Origin and Assessment of Cotton Fiber Maturity*, International Institute for Cotton, Manchester, England, 1988.

567. Pierce, F.T. and Lord, E., *J. Text. Inst.*, 30, T173, 1939.

568. Raes, G.T.J. and Verschraege, L., *J. Text. Inst.*, 72, 191, 1981.

569. Standard Test Method for Maturity of Cotton Fibers (Sodium Hydroxide Swelling and Polarized Light Procedures), Vol. 07.01, ASTM Designation D1442-00, December 2000.

570. Smith, J.C., McCracken, F.L., Schiefer, H.F., and Stone, W.K., *Text. Res. J.*, 26, 281, 1956.

571. Calkins, E.W.S., *Text. Res. J.*, 6, 441, 1946.

572. Thibodeaux, D.P. and Evans, J.P., *Text. Res. J.*, 56, 130, 1986.

573. Boylston, E.K., Thibodeaux, D.P., and Evans, J.P., *Text. Res. J.*, 60, 80, 1993.

574. Thibodeaux, D.P. and Price, J.B., *Melliand Textilber.*, 70, 243, 1989.

575. Lunenschloss, J., Gilhaus, K., and Hoffman, K., *Melliand Textilber.*, 61, 5, 1980.

576. Hertel, K.L. and Craven, C.J., *Text. Res. J.*, 21, 765, 1951.

577. Ghosh, S., *Text. World*, 28, 45, 1985.
578. Montalvo, J.G., Faught, S.E., Buco, S.M., Saxton, and A.M., *Appl. Spectrosc.*, 41, 645, 1987.
579. Linear Density of Cotton Fibers (Array Sample), ASTM Designation D1769–77, 1982.
580. Lord, E., *J. Text. Inst.*, 47, T16, 1956.
581. Morton, W.E. and Hearle, J.W.S., *Physical Properties of Textile Fibers*, John Wiley & Sons, New York, 1976.
582. Meredith, R., *J. Text. Inst.*, 36, T107, 1945.
583. Hearle, J.W.S., *J. Appl. Polym. Sci., Appl. Polym.Symp.*, 47, 1, 1991.
584. Hebert, J.J., Thibodeaux, D.P., Shofner, F.M., Singletary, J.K., and Patelke, D.B., *Text. Res. J.*, 65, 440, 1995.
585. Lord, E., *Manual of Cotton Spinning, Vol. II, part 1—The Characteristics of Raw Cotton*, Textile Book Publishers, Inc., New York, 1961, p. 214.
586. Meredith, R., *J. Text. Inst.*, 37, T205, 1946.
587. Alexander, E., Lewin, M., Musham, H.V., and Shiloh, M., *Text. Res. J.*, 26, 606, 1956.
588. Hearle, J.W.S., *Text. Res. J.*, 24, 307, 1954.
589. Walker, A.C. and Quell, M.H., *J. Text. Inst.*, 24, T123, 1933.
590. Hearle, J.W.S., *J. Text. Inst.*, 44, T117, 1953.
591. Davidonis, G.H., Johnson, A., Landivar, J.A., and Hood, K.B., *Text. Res. J.*, 69, 754, 1999.
592. Zurek, W., Greszta, M., and Frydrych, I., *Text. Res. J.*, 69, 804, 1999.
593. Krowicki, R.S., Hinojosa, O., Thibodeaux, D.P., and Duckett, K.E., *Text. Res. J.*, 66, 70, 1996.
594. Doberczak, A., Dowgielewicz, St., and Zurek, W., *Cotton, Bast, and Wool Fibers*, (translated from Polish) published for U.S. Department. of Agriculture and the National Science Foundation, Washington, D.C. by Centralny Instytut Informacji Naukowo-Techniczncj I Ekonomiczncj, Warszawa, Poland, 1964, pp. 16–160.
595. Hunter, L., *Textiles: Some Technical Information and Data V: Cotton*, South Africa Wool and Textile Research Institute Special Publication, Port Elizabeth, Republic of South Africa, 1981.
596. Cotton, *Technical Monograph*, No. 3, Ciba–Geigy Agrochemicals, Ciba–Geigy, Ltd., Basel, Switzerland, 1972.
597. Elliot, F.C., Hoover, M., and Porter, W.K., Jr., *Cotton: Principles and Practices*, The Iowa State University Press, Ames, IA, 1968.
598. Reeves, B.G. and Garner, W.E., *CRC Handbook of Transportation and Marketing in Agriculture*, Vol. II, *Field Crops*, The Chemical Rubber Company., Cleveland, OH, 1982, pp. 353–384.
599. Jordan, A.G. and Needham, D.K., Cotton bale weight, density and size standardization, Proceedings of the International Cottontest-Conference, Bremen, Faserinstitut Bremen eV, Bremen, West Germany, 1982.
600. Hinojosa, O. and Thibodeaux, D.P., *Text. Chem. Color.*, 25(1), 27, 1993.
601. The Classification of Cotton, in *Agricultural Handbook Number*, 566, Agricultural Marketing Service, U.S. Department. of Agriculture, Washington, D.C., 1993.
602. Anon., Cotton management in China, Indian Cotton Mill Federation, June 1989, pp. 79–81.
603. Ustyugin, V.E., Development of standards and certification of cotton fibre in Uzbekistan, Fourth International Cotton Conference, Gdynia, Poland, September 14–15, 1995, pp. 33–40.
604. Macdonald, A.G., Expert panel on commercial standardization of instrument testing of cotton (CSITC), Proceedings of the International Cotton Conference, Bremen, Faserinstitut Bremen eV, 2004.
605. Belhia, A., The Cotton industry in Azerbaijan, *Cotton International*, 1992, p. 192.
606. Personal communication, Eric Hequet, Head Cotton Technology Laboratory, Centre de Cooperation Internationale en Recherche Agronomique Pour le Development (CIRAD), Paris, France, 1995.
607. Mursal, I.E., Field classification restores Sudan cotton quality, *Cotton International*, 1994, p. 219.
608. Personal communication, Hein Schroder, Quality Control Manager, South Africa Cotton Board, 1995.
609. Assal, A., and El, S., Adoption of new instrumental grading system for Pakistan's cotton, Proceedings of the Internatonal Committee on Cotton Testing Methods, Bremen, Germany, March 1–2, 1994.

610. *Natural Threads, The Australian Cotton Story*, Australian Cotton Foundation, Ltd., Waterloo, Australia, 1993.
611. Hunter, L., Proceedings of the 27th International Cotton Conference, Bremen, Faserinstitut, Bremen, eV, 2004, pp. 62–70.
612. Suh, M.W. and Sasser, P.E. *J. Text. Inst.*, 87(3), 43, 1996.
613. Townsend, T.P., Commercial standardization of instrument testing of cotton: How soon a reality?, Proceedings of the 2005 Beltwide Cotton Conference, National Cotton Council, Memphis, TN, 2005, pp. 2386–2389.
614. Adams, G., Slinsky, S., Boyd, S., and Huffman, M., *The Economic Outlook for U.S. Cotton*, 1996 edition, Economic Services, National Cotton Council of America, Memphis, TN, 2005.
615. Meyer, L., MacDonald, S., and Skinner, R., Cotton and wool outlook, USDA, CWS-05e, June 13, 2005 [Electronic Outlook Report from the Economic Research Service, www.ers.usda.gov].
616. Anon., *Cottonseed and Its Products*, 9th ed., National Cottonseed Products Association, Memphis, TN, 1990; O'Brien, R.D., Jones, L.A., King, C.C., Wakelyn, P.J., and Wan, P.J., Cottonseed Oil, in *Bailey's Industrial Oil and Fat Products*, 6th ed., Vol. 2, Shadida, F., Ed., John Wiley & Sons, New York, 2005, chap. 5, pp. 173–279.
617. Huffman, M., *Cotton Counts Its Customers (The Quality of Cotton Consumed in Final Uses Produced in the United States)*, 2004 edition, National Cotton Council of America, Memphis, TN, 2004.
618. Anon., *Cotton Dyeing and Finishing: A Technical Guide*, Cotton Incorporated, Cary, NC, 1996.
619. Scheurell, D.M., Spivak, S.M., and Hollies, N.R.S., *Text. Res. J.*, 55, 394, 1985.
620. Choi, H., Moreau, J.P., and Srinivasan, M., *J. Environ. Sci. Health*, A29(10), 2151, 1994.
621. Choi, H. and Cloud, R.M., *Environ. Sci. Technol.*, 26, 772, 1992.
622. *Fibers as Renewal Resources for Industrial Materials*, National Academy of Sciences, Washington, D.C., 1976.
623. *Renewal Resources for Industrial Materials*, National Academy of Sciences, Washington, D.C., 1976.
624. *Long-Term Agricultural Baseline Projections 1995–2005*, U.S. Department of Agriculture, World Agricultural Outlook Board, Staff Report, WAOB-95-1, 1995.
625. Robinson, E., Cotton Analysts to Come out Swinging in July, Delta Farm Press, June 29, 2005.
626. Strahl, W.A., Cotton in rugs and carpets, Proceedings of the 1996 Beltwide Cotton Research Conference, National Cotton Council, Memphis, TN, 1996, pp. 1501–1507.
627. Wakelyn, P.J., Menchey, K., and Jordan, A.G., Cotton and environmental issues, in *Cotton—Global Challenges and the Future*, Papers Presented at a Technical Seminar at the 59th Plenary Meeting of the International Cotton Advisory Committee (ICAC), Cairns, Australia, November 9, 2000, pp. 3–11.
628. King, E.G., Phillips, J.R., and Coleman, R.J., Eds., *Cotton Insects and Mites: Characterization and Management*, The Cotton Foundation Reference Book Series, No. 3, The Cotton Foundation, Memphis, TN, 1996.
629. McWorter, C.G. and Abernathy, J.R., Eds., *Weeds of Cotton: Characterization and Control*, The Cotton Foundation Reference Book Series, No. 2, The Cotton Foundation, Memphis, TN, 1996.
630. Kirkpatrick, T.L. and Rothrock, C.S., Eds., *Compendium of Cotton Diseases*, 2nd ed., APS Press, St. Paul, MN, 2001.
631. *Cotton Nematodes, Your Hidden Enemies: Identification and Control*, The Cotton Foundation, National Cotton Council, and Aventis Crop Science, 2002.
632. Cotty, P.J., Cottonseed losses and mycotoxins, in *Compendium of Cotton Diseases*, 2nd ed., Kirkpatrick, T.L. and Rothrock, C.S., Eds., APS Press, St. Paul, MN, 2001, pp. 9–13.
633. Cotty, P.J., *Phytopathology*, 84, 1270, 1994.
634. Castellan, R.M., Olenchock, S.A., Kingsley, K.B., and Hankinson, J.L., *N. Engl. J. Med.*, 317, 605, 1987.
635. Jacobs, R.R. and Wakelyn, P.J., Assessment of toxicology (respiratory risk) associated with airborne fibrous cotton-related dust, Proceedings of the 1998 Beltwide Cotton Conference, National Cotton Council, Memphis, TN, 1998, pp. 213–220.
636. Cotton dust, U.S. Code of Federal Regulations, 29CFR1910.1043, (a) Scope and application and (c) Permissible exposure limits and action levels.

637. Wakelyn, P.J., Greenblatt, G.A., Brown, D.F., and Tripp, V.W., *Am. Ind. Hyg. Assoc. J.*, 37, 22, 1976.
638. Pickering, C.A.C., The search for the aetiological agent and pathogenic mechanisms of byssinosis: a clinician's view of byssinosis, Proceedings of the 15th Cotton Dust Research Conference, Jacobs, R.R., Wakelyn, P.J., and Domelsmith, L.N., Eds., National Cotton Council, Memphis, TN, 1991, pp. 298–299.
639. Rohrbach, M.S., The search for the aetiological agent and pathogenic mechanisms of byssinosis: a review of *in vitro* studies, Proceedings of the 15th Cotton Dust Research Conference, Jacobs, R.R., Wakelyn, P.J., and Domelsmith, L.N., Eds., National Cotton Council, Memphis, TN, 1991, pp. 300–306.
640. Nichols, P.J., The search for the aetiological agent and pathogenic mechanisms of byssinosis: *in vivo* studies, Proceedings of the 15th Cotton Dust Research Conference, Jacobs, R.R., Wakelyn, P.J., and Domelsmith, L.N., Eds., National Cotton Council, Memphis, TN, 1991, pp. 307–320.
641. Glindmeyer, H.W., Lefants, J.J., Jones, R.N., Rando, R.J., Kader, H.N.A., and Weill, H., *Am. Rev. Respir. Dis.*, 144, 675, 1991.
642. Perkins, H.H., Jr., and Olenchock, S.A., *Annu. Agric. Environ. Med.*, 2, 1, 1995.
643. The Task Force for Byssinosis Prevention, Washed cotton, a review and recommendations regarding batch kier washed cotton, *Current Intelligence Bulletin*, 56, U.S. Department of Health and Human Service, NIOSH, Aug. 1995.
644. Occupational exposure to cotton dust, Final Rule, Federal Register, 50, pp. 51120–51179, December 13, 1985; Washed cotton, 29 CFR 1910.1043(n); Occupational exposure to cotton dust, Direct Final Rule, Federal Register, 65, pp. 76563–76567, December 7, 2000.
645. Formaldehyde, U.S. Code of Federal Regulations, 29 CFR 1910.1048.
646. Assessment and control of indoor air pollution, Report to Congress on Indoor Air Quality, Vol. 1, Office of Air and Radiation, U.S. Environmental Protection Agency, 1989; Formaldehyde risk-assessment update, Office of Toxic Substances, U.S. Environmental Protection Agency, Washington, D.C., June 11, 1991.
647. IARC classifies formaldehyde as carcinogenic to humans, International Agency for Research on Cancer, June 2004.
648. Robbins, J.D., Norred, W.P., Bathija, A., and Ulsamer, A.G., *J. Toxicol. Environ. Health*, 14, 453, 1984.
649. Robins, J.D. and Norred, W.P., Bioavailability in Rabbits of Formaldehyde from Durable Press Textiles, Final Report on CPSC IAG 80-1397, USDA Toxicology and Biological Constituents Research Unit, Athens, GA, 1984.
650. *Status Report on Formaldehyde in Textiles Portion of Dyes and Finishes Project*, U.S. Consumer Product Safety Commission, Washington, D.C., January 3, 1984.

Index